Political Responsibility for Climate Change

This book offers new perspectives on how social and political institutions can respond more effectively to climate change.

Theresa Scavenius presents a concept of moral responsibility that does not address the obligations of individual citizens, but instead assesses the moral responsibility of institutionalised actors, such as governments, parliaments, and other governmental agencies. This focus on political responsibility is something that up until now has largely been neglected by moral theory, but Scavenius argues in this book that accountability must be assigned to institutionalised group agents. With this new research, she outlines building blocks for a new agenda of climate studies by offering an innovative approach to climate governance and democratic climate action at a time when many political initiatives have failed and crucially outlines the necessity of approaching moral dilemmas from a fact-sensitive political theoretical approach.

Written in a clear and engaging style, this volume will be an invaluable reference for researchers interested in moral philosophy, climate change, environmental politics and policy, and institutional theory.

Theresa Scavenius is an Associate Professor in the Department of Planning, University of Aalborg, Copenhagen, Denmark.

Routledge Advances in Climate Change Research

Climate Change, Moral Panics and Civilization
Amanda Rohloff
Edited by André Saramago

Climate Change and Social Inequality
The Health and Social Costs of Global Warming
Merrill Singer

Cities Leading Climate Action
Urban Policy and Planning
Sabrina Dekker

Culture, Space and Climate Change
Vulnerability and Resilience in European Coastal Areas
Thorsten Heimann

Communication Strategies for Engaging Climate Skeptics
Religion and the Environment
Emma Frances Bloomfield

Regenerative Urban Development, Climate Change and the Common Good
Edited by Beth Schaefer Caniglia, Beatrice Frank, John L. Knott Jr., Kenneth S. Sagendorf, Eugene A. Wilkerson

Local Activism for Global Climate Justice
The Great Lakes Watershed
Edited by Patricia E. Perkins

Political Responsibility for Climate Change
Ethical Institutions and Fact-Sensitive Theory
Theresa Scavenius

For more information about this series, please visit: www.routledge.com/Routledge-Advances-in-Climate-Change-Research/book-series/RACCR

Political Responsibility for Climate Change

Ethical Institutions and Fact-Sensitive Theory

Theresa Scavenius

LONDON AND NEW YORK

First published 2020
by Routledge
2 Park Square, Milton Park, Abingdon, Oxon OX14 4RN

and by Routledge
605 Third Avenue, New York, NY 10017

First issued in paperback 2021

Routledge is an imprint of the Taylor & Francis Group, an informa business

© 2020 Theresa Scavenius

British Library Cataloguing-in-Publication Data
A catalogue record for this book is available from the British Library

Library of Congress Cataloging-in-Publication Data
A catalog record for this book has been requested

ISBN 13: 978-0-367-78468-3 (pbk)
ISBN 13: 978-0-367-18970-9 (hbk)

Typeset in Times New Roman
by Apex CoVantage, LLC

In memoriam
Merle Scavenius

Contents

List of tables   viii

Introduction   1

PART 1
Fact-sensitivity and normativity   19

1   Fact-sensitive political theory   21
2   The indeterminacy challenge   32
3   Fact-sensitive ought-assignments   48

PART 2
Fitness-conditions of moral responsibility   71

4   Fitness-conditions of rational agency   73
5   Fitness-conditions of group agents   95
6   Control conditions and democratic climate governance   115

PART 3
Moral responsibility for climate change   131

7   Collective responsibility   133
8   Moral excuse and democratic citizens   141
9   Collective responsibility and democratic institutions   157
    Conclusion   176

Index   180

Tables

1.1   Four types of normativity   22
3.1   Four types of constraints   54
4.1   Snatch dilemma I   78
4.2   Snatch dilemma II – modified version   79

Introduction

This book concerns what type of agents *can* be held responsible and what type of agents *should* be held responsible for anthropogenic climate change. I offer a *general* theory of why can- and fitness-conditions of moral agency are necessary, albeit not sufficient, premises for the assignment of ought to particular agents. I also provide a special theory of why institutionalised group agents should be held morally responsible for climate change. The objective of this work is to combine insights from moral philosophy and descriptive political science; the account of moral responsibility for climate change offered is theoretical and normative in character.

The political failures of climate actions

There is a vast accumulation of technical knowledge about climate change. For a decade now, political and social scientists have documented in great detail how global climate change will have severe impacts on social, human, and environmental conditions. Human choices, industrial production, and lifestyles affect the dynamics of the planet's climate and the basic life-support system for all forms of life (N. Stern 2006; Gardiner 2010).

The need to integrate environmental considerations into national socio-economic planning is now widely recognised (Rose 2011). Advanced environmental legislation and climate strategies are implemented in many cities, municipalities, and countries across the world (Bulkeley 2013; Hoff and Strobel 2013). In 2011, Rose counted 700 international environmental agreements that exist at bilateral, regional, and multilateral levels (Rose 2011). Today, there are many more.

Although many citizens, stakeholders, and nation-states have made great efforts in initiating policies to manage both climate change and the unsustainability of natural resource consumption, two statistics expose the profound political challenges that remain. First, the total consumption of coal, considered to be critically damaging to human health and the environment, rose 50 percent in the decade between 2000 and 2010 (IEA 2011). Second, in 2013, the concentration of carbon dioxide in the atmosphere reached 400 parts per million (ppm) – its highest level since measurements and collection of data began in 1956. Since then, the concentration has increased every year. Yet despite the good intentions and promising results of many of these global climate initiatives, these numbers tell us a sobering truth: we have yet to manage and control the combustion of fossil fuels. As if this weren't enough, according to a recent World Wildlife Fund study, most of the earth's natural resources are overused and likely to collapse in the near future (WWF 2012).

At the global level, the United Nations has hosted several international climate conventions since 1992, and the International Panel for Climate Change (IPCC) has collected an ever-increasing body of scientific evidence on climate change. The two main international agreements on environmental and climate policy are the 1987 Montreal Protocol, which aims to reduce the emissions of chlorofluoro-carbons (CFCs) and other ozone-depleting substances, and the 1997 Kyoto Pro-tocol, which aims to reduce the emissions of greenhouse gases (GHG).[1] Whilst the Montreal Protocol is widely recognised as having been successful in reducing CFCs, the Kyoto Protocol has failed to reduce GHGs significantly (Prins and Rayner 2006).[2]

Another political failure was the Conference of the Parties (COP)15 meeting in Copenhagen in 2009, which in international academic and policy communi-ties has become a point of reference as a great disappointment of contemporary global climate policy (Gardiner 2011).[3] In Copenhagen, the world's leaders failed to agree on a legally binding global agreement. This fiasco is commonly held as proof of the lack of willingness to pay the costs of re-establishing an environmen-tally sustainable economy (Hoff and Strobel 2013). The Paris Agreement, ratified in 2015, is also a non-binding agreement.

Climate responsibility

An important debate now underway among climate change advocates, scholars, and theorists concerns the level of government where the bulk of climate change policy and politics should be implemented. Some scholars are concerned about how to mitigate CO_2 emissions at the global level. From this perspective, the failure of the COP meetings to produce a legally binding global treatment was a great misfortune that generated tremendous political defeatism in the international climate change community. In contrast, other scholars are concerned with how *local* levels of governance can effectively contribute to climate mitigation and natural resource management (ICLEI 2010). For example, there is great optimism at the local level for transferring knowledge between cities across the globe and tackling the issue of how to coordinate multi-stakeholder governance in which civil society, political-administrative institutions, and private corporations work jointly to take climate governance to new levels.

Theoretically, there exists no adequate set of theoretical concepts that allows us to comprehend this recent development in local climate governance. This theoret-ical gap is caused by two scholarly focuses: studies in climate ethics and politics tend to focus on either the economic aspects of mitigation or on the moral claims for global justice. I argue that neither of these approaches provides a satisfac-tory account of how it is possible to govern climate change politically. Moreover, traditional scholarly approaches tend to underestimate what political societies are capable of accomplishing. In effect, what is needed is to overcome the prisoner's dilemma challenges, which economists have difficulty solving, and to overcome the dichotomy between local and global justice, which many moral philosophers claim is difficult to bridge.

A dominant approach to climate responsibility is the *common, but differenti-ated responsibility*, which is reflected, for example, in the distinction between Annex I and II countries in the Kyoto Protocol. Common but differentiated responsibility means that every country has a responsibility, but that rich coun-tries should bear the costs of mitigating climate change. However, one example of the conundrum that the issue of common but differentiated responsibility raises is the issue of China and India. Historically, and if counted in average per-capita emissions, China and India belong to the group of poor, low-emission countries (Hardin 2012, 128). However, if counted in total numbers, China emits a higher amount of CO_2 than the United States (US). For example, there is a growing class of super-rich people in low- and middle-income countries with disproportionately large environmental footprints (Hardin 2012, 123). Thus, the distinction between rich and poor countries has become a huge obstacle for global climate negotia-tions, which is made even more intractable because the US will not accept that it must bear a heavier burden than, say, China and India.

In order to comprehend these recent trends in how global inequality resembles national inequality, we need a new framework for assigning moral responsibility to national and regional governmental authorities. This allocation of responsi-bility is necessary not merely with regard to the global contributions to climate change because of the environmental harm and carbon pollution associated with climate change and the current lack of local and national governance of natu-ral resource maintenance. This should not be perceived as an argument for not holding these governmental authorities responsible for their engagement in global politics, nor does it mean that other agents, such as private multinational cor-porations and international organisations, should not be assigned responsibility. The account I defend concerns merely one conceptualisation of a fully elaborated theory of moral responsibility for climate change, which is an important but fre-quently neglected topic within climate studies.

More specifically, I shall draw attention to the responsibilities which national and regional entities have as institutionalised group agents. In global environmen-tal studies, the democratic level of politics, which is capable of managing envi-ronmental harm and climate change, has to some extent been neglected because of the dominant dichotomist understanding of the global versus national level of politics. In the global justice literature, this dichotomist approach has caused a lack of interest in considering what national political institutions should be held responsible for.

I argue that the dominant economic and moral approaches to climate ethics and politics have paralysed the conceptualisation of how political institutions are fit to govern climate change by confining the debate to questions about costs and emis-sions. A purely economic approach to climate change wrongly conceptualises the challenge of climate change and implies policy solutions that are incapable of addressing the underlying causes of climate change. I argue that the focus on the mitigation of emissions does not help us understand the intermediating societal landscape of agency, rationality, and politics that constitute the behavioural space within which climate change is happening (Fukuyama 2012; Stehr 2008; Plant 2009; Streeck 2011; Prins and Caine 2013).

The analysis I present here reconstructs a policy framework with a renewed focus on how social and institutional capacities *can* and *ought* to process the political challenges of climate change. This work establishes the foundation for a revised picture of climate politics and ethics by offering a theory of how institutionalised collective agents ought to be held morally responsible for the current lack of environmental and climate governance. I offer three strategies in the development of this argument: a fact-sensitive account of normativity and moral responsibility (Part 1), a discussion of which methodological theories could provide a framework in which individuals and institutionalised group agents are *fit* to be held morally responsible for climate change (Part 2), and a normative discussion of who *ought* to bear responsibility for not doing enough with regard to climate governance and for what normative reasons (Part 3).

Moral responsibility

Moral theories provide us reasons to criticise acts and policies that fail to avoid environmental harm and climate change. Two reasons, a human-centric instrumental reason and an eco-centric intrinsic reason, can be distinguished.[4] The human-centric approach argues that the earth's natural bio-capacity has instrumental value for upholding the well-being of humans and their rights to life, health, and basic subsistence (Nagel 1995; Shue 1999; Caney 2010a). If climate change and other kinds of environmental harm deprive humans of their basic freedoms and capabilities, it is wrong to harm the climate and environment.

The eco-centric argument defends the preservation of the biosphere 'for its own sake, irrespective of the possible benefits to humans that might flow from so doing' (Singer 2010, 280). Most human-centric arguments are based on a commitment to the inviolability of human rights, whilst the eco-centric argument stresses the moral significance of all sentient creatures – human and non-human, present and future – and the physical and biological conditions under which they live (Singer 2010, 284–288). In this approach, behaviour and actions which have a negative impact on the planet's climate should be avoided.[5] The eco-centric argument ascribes non-instrumental value to nature, animals, and the ecosystem. Proponents of the eco-centric position disagree with proponents of the human-centric position as to what extent concerns for non-human bodies can trump concerns for human bodies (Hassoun 2011, 237). Despite this, both positions provide reasons for why it is morally impermissible to contribute to environmental harm and climate change.

Notwithstanding the clear moral wrongdoing with regard to climate change, the theory of who ought to be held morally responsible for climate change is not yet adequately conceptualised. Crucially, the dominant theory of moral wrongdoing does not address to whom it is fair to assign moral responsibility for climate change. There is an important connection between a moral philosophy of *what and who have been wronged* and a political theory of *to whom it is fair to assign moral responsibility*: if the political theory bears no relationship to the moral issues, then the political theory would be redundant. Nevertheless, the political theory of to whom it is fair to assign moral responsibility does not collapse into the moral theory of what and who has been wronged.

There are different ways of conceptualising moral responsibility. Recent contributions to the discussion distinguish two aspects: 1) *causal* or *outcome responsibility*, where agent A is responsible for X when A has caused X and 2) *moral blame*, where agent A is morally responsible for X when A is blameworthy or praiseworthy for doing or failing to do X (Eshleman 2009; Knight and Stemplowska 2011, 12).[6] It is common for the different theories of moral responsibility to combine a set of *normative* properties with a set of *factual* ones.

Consider, for example, this combination: agent A is responsible for an action if A has knowledge about the action's condemnable outcome (O) but, nonetheless, performs the action. Suppose A drives down a narrow street at a high speed. In this situation, A should know that the action 'to drive down a narrow street at a high speed' could have outcome O: 'kills someone'. Assuming that A knows or ought to have known the possible outcome of her action, A is responsible for the condemnable consequences.

This example resembles relevant dimensions of what we could call *simple* cases of moral responsibility. By simple, I mean that because there is a clear factual aspect establishing the causal link between the agent (A) and the outcome (O), a clear normative criterion in this case might be if an agent is aware of an act's possible bad outcome, the agent ought not to pursue the act. Taking this argument as the point of departure, some theorists focus on the CO_2 emitted by cars and argue that the responsible agents for climate change are the car drivers.

In contrast, I argue that climate change is a case of *complex* moral responsibility, where we cannot assume without further justification that the relevant moral agent is the individual car driver. In the present context, I take complex cases to be cases where the total (bad) outcome is scientifically well documented,[7] and the moral wrongdoing is clear, but where the question of *who ought to do what* can be answered neither by the scientific data nor by the moral theories alone.

As far as theories of moral ought-assignments are concerned, these are typically causal theories framed in terms of descriptive and normative models. Although such theories contribute to our understanding of climate ethics by describing aspects of the underlying social mechanisms which are causing climate change, they do not necessarily enable us to assign particular agents the moral responsibility for climate change.

To do so requires an elaborated account of moral responsibility that unpacks both the factual and normative aspects of ought-assignments. Whatever the situation is in climate science and moral philosophy, when one considers the political theoretical question of to whom it is fair to assign moral responsibility for climate change, the methodological and sociological assumptions of agency, causality, and rationality need to be clarified. Several methodologies and sociological theories compete with each other when explaining social events and establishing causal relationships between conditions, acts, and outcomes. As a result, the theory of moral responsibility for climate change is susceptible to its methodological and sociological assumptions.

Hence, the point of departure for this work is that the question of which agents are morally responsible for the failures of climate action is not merely a question of normative criteria but also one of sociological theory about agent, causality, and rationality. Framing the discussion in this way makes it clear that we need more flesh on the aforementioned definition of moral responsibility and the particular mechanisms by which moral responsibility is constituted by both factual and normative components. Since the purpose of this book is to develop a theory of moral responsibility for climate change, we need to outline the working assumption of moral responsibility by stressing three preliminary points.

First, the level of relevant moral agency cannot be assumed. Second, the causality and explanation cannot be intuitively established. A solid methodological theory is needed that is capable of establishing a valid causal relation and explanation between the agent and the outcome. Third, a set of normative criteria is needed that provides a sound reason to assign moral responsibility to particular types of agents. The reason for the last point is that we cannot assume that the same normative criteria apply for both lower and higher levels of agency. I argue that the normative criteria for climate responsibility differ between individual and group agents, such as democratic political institutions. Thus, a preliminary account of moral responsibility for climate change can be defined as *A is morally responsible for causing X*:

1   If and only if A is a moral agent;
2   If and only if there is a plausible and reasonable explanatory theory which clarifies why A caused X; and
3   If and only if there is a normatively warranted and sound reason for assigning responsibility to A for X.

In light of this formula, we can now see that moral responsibility consists of (1) a *descriptive* component, which investigates the factual criteria for agency, causality, and explanation and (2) a *normative* component concerning the normative criteria for who we should regard as morally responsible and who we should blame for failing to avoid what is morally impermissible. In this context, the purpose here is to develop a better and more coherent account of moral responsibility for climate change. Part 1 proposes and defends a fact-sensitive account of moral responsibility, which demonstrates how descriptive and normative aspects are two distinct, albeit equally important, components of moral responsibility. In this section, I elaborate on the meta-theoretical account of how to incorporate the factual and sociological level of analysis into the normative theory without losing the validity and soundness of the latter.

Part 2 discusses under what conditions individual agents and group agents *can* and are *fit* to be held responsible for climate change. I argue that many current theories neglect to consider that some of the fitness-conditions for moral responsibility are framed by explanatory theories and methodological assumptions regarding agency. With this in mind, I discuss in Part 3 the normative criteria for when individual and group agents *ought* to be held responsible.

Idealism and realism

Contemporary debates on moral responsibility for climate change have difficulty capturing moral dilemmas that contain both factual and normative considerations. One reason for this difficulty is that the discussion is torn between proponents of ideal theory and realist theory, who situate the locus of factual and normative aspects at opposite ends of the spectrum of moral theory.[8] The unfortunate result of this split is that each end of the spectrum is scrutinised thoroughly whilst little attention is given to the intersections between the two.

In this debate, proponents of idealism focus on the pure normative aspects of responsibility.[9] Of course, this does not imply that they deny the relevance of factual aspects: in many cases, they assume several factual assumptions such as rationality and cognitive and bodily capacities. But idealists frequently under-theorise these factual assumptions because they are assumed to be of little relevance for the normative inquiry. Instead, they examine the status and content of normative principles without addressing the potential practical implications (O'Neill 2013, 3). With regard to moral responsibility for climate change, two types of critique of the idealist claims should be noted. One, idealists risk being *too demanding* and thereby risk losing the chance to be action-guiding (which we typically want moral theories to be). Second, they underdetermine concrete obligations and the so-called imperfect duties.

At the other end of the spectrum, proponents of realism focus on the factual aspects of and constraints to the assignment of responsibility and other types of obligations. There is a lengthy tradition in political theory and international relations studies which upholds a realistic understanding of political societies.[10] The realists presume a sceptical anthropology of human willingness and capacity to engage in (ideal) political interactions. Proponents of realism frequently criticise idealism for being an 'empty formalism' without any political relevance whose concept of responsibility is too demanding and lacks practical implications (O'Neill 2013, 34). The realist, on the other hand, is frequently criticised for sliding towards moral relativism by neglecting and belittling trans-cultural and trans-historical normativity (Lægaard 2006, 412). One reason for this is that the realist takes an overly strict approach to the factual circumstances of politics. For example, Bernard Williams argued,

> Political theory [. . .] must start from and be concerned in the first instance not with how people ought ideally (or ought 'rationally') to act, what they ought to desire, or value, the kind of people they ought to be, etc., but, rather, with

the way the social, economic, political, etc., institutions *actually* operate in some society at some given time, and what *really* does move human being to act in given circumstances.

(Geuss 2008, 9, emphasis added)

Here, there is a presumption of a static model for what institutions 'actually are' and what 'really moves' individuals. For reasons that will become clear later, the static model is circumstanced by those who point out that the conditions for institutions and agency may change. Sometimes this happens gradually; sometimes this happens in revolutionary movements. Arguably, individuals' motivations, choices, and rationality are more context independent, in contrast to what the realists tend to accept (Le Grand 2006).

Currently, many scholars defend positions which fall between idealism and realism. One objective in defining these intermediate positions is to provide an account that allows for a political theory that is neither too ideal nor too remote from actual politics and current circumstances of politics. Contrary to the realists, who question the existence of universal moral principles, the non-ideal realist accepts the validity of universal moral principles but suggests a set of non-ideal normative principle that is more suitable given the current 'reality', which includes 'power', 'state interest', the 'non-ideal circumstances of justice', and 'people's capacity and interest', all of which are thought to limit the feasibility of implementing universal moral principles (Williams 2005). However, proponents of the non-idealist position risk becoming *too unambitious*. In other words, they risk letting too many people off the hook with regard to responsibility for climate change if they accept people's *non-ideal* reasons not to act in an environmentally friendly way.

Fact-sensitivity and normativity

Whilst it might be correct (as the realists argue) that political theory should not be too remote from politics, I argue that the opposite is also correct: political theory should not be too *realistic* and too close to actual politics to the extent that trans-historical and trans-cultural norms and ideals are neglected or rejected. If robust and sound normative principles are rejected, the argument for moral responsibility will be incomplete, since the normative justification is impossible to warrant. Moreover, if we do not allow for strict normative standards by which contemporary failures of climate action can be scrutinised, the discussion of moral responsibility becomes redundant or very unambitious. On the other hand, if the political theoretical discussions of climate ethics are too idealistic, there is a risk that the principles will become too demanding, thereby defeating the purpose of suggesting an action-guiding theory of moral responsibility. Instead, what we want is a third position that allows for *ambitious ought-assignments that are not too demanding*.

The goal of this analysis is to develop such a framework, which I call a *fact-sensitive normative theory of moral responsibility*. This account has the same objective as the realist theory: to provide a politically relevant approach with the potential to guide real-world politics of global challenges. Fact-sensitive political theory, however, differs from the realist approaches by maintaining a robust and sound normativity. But how is that possible? Is it like squaring the circle? I argue that such an approach is possible if several meta-theoretical arguments are accepted. The approach to fact-sensitivity is developed through discussions of contemporary contributions to idealism and realism, especially in the global justice literature.

One of the primary arguments for this approach is that we should not relax the normative criteria, but instead open the black box of agency, rationality, and explanation. If we look closer at this analysis, new types of agents and new ways of understanding the problems of climate change are revealed. My account of moral responsibility is, therefore, not a discussion of realism and non-idealism versus idealism. Its objective is to establish a different axis that distinguishes between *abstract* normative reasoning and *concrete* normative reasoning. As Onora O'Neill has argued, many confuse *abstract* normative reasoning with *ideal* normative reasoning and *realist* normative with *concrete* normative reasoning (O'Neill 1988, 11–12).

The merits of abstract versus non-abstract reasoning are that the normative debate is not torn between two distinct ends of a spectrum. Instead, the locus of the investigations is vertical in the sense that abstract and concrete normative discussions are given equal status: they can co-exist compatibly. One of the primary differences between the abstract and concrete normative principles, which is discussed in greater detail in Chapter 1, is whether abstract or concrete facts are considered necessary for the determination of normative principles. Thus, the fact-sensitive account of normativity is neither realist nor non-ideal, but an ambitious and demanding normative theory which contains both abstract and concrete normative reasoning.

A number of theories have developed their favoured basic accounts of what I call 'concrete' normative principles. This includes theories of non-idealism (Rawls 1999), applied ethics (Singer 1993), and feasibility constraints (Sen 2009). These scholars have developed accounts of how to do political philosophy in light of factual constraints. The kind of fact-sensitivity I defend takes a different point of departure by asking how to develop a concrete and policy-specific political theory in the light of fundamental moral principles. Ultimately, there will be several theoretical overlaps between the accounts of fact-sensitivity, applied ethics, and feasibility studies. But by asking a different set of questions, the conceptualisation of fact-sensitivity contributes to the current theoretical debate by perceiving factual components not as descriptive elements but as methodologically embedded theoretical assumptions of agency and rationality that have normative implications. By so arguing, I allow for a closer theoretical link between the conceptualisation and theorisation of facts and norms (Miller 2010).

By contrast, many moral philosophers and empirical scientists argue that studies of facts should be conducted independently of studies of normativity. One reason for this argument is that empirical scientists fear a normative bias, whilst the moral philosophers fear that the contingency of facts will undermine the validity of normative inquiries (Brown 1977, 220; Barber 2006, 541). Independency between facts and norms is closely associated with the principle that rejects the possibility of deriving a normative conclusion from a factual premise. This principle is commonly coined as the naturalistic fallacy, which says that we cannot derive *it is good that p* from *that p*. In order to make a normative conclusion, at least one normative premise is required.

This interpretation, however, does not capture the full complexity of the relationship between facts and norms and the normative challenges of ought-assignment and moral responsibility with which I am occupied in this work. Here, the normative conclusions contain a combination of normative premises about what is just, or fair, together with factual premises about agency, social norms, and rationality. This combination becomes particularly apparent in cases of *can* and *fitness*-conditions which are related to ought-judgements. Here, we are not interested in who *can* do *what* in an empirical sense. Instead, we are interested in the theoretical and methodological justification of *who* is fit to be held morally responsible for climate change. The fact-sensitive account identifies the *can*- and *fitness*-conditions of ought-judgements as concrete fact-sensitive principles that contain theoretical assumptions about agency, rationality, and norms. In order to be justified, the fact-sensitive principles should cohere with some more fundamental principles of justice and moral responsibility.

Outline

The book is organised into three parts. The objective of Part 1 is twofold. It establishes the foundation for a framework that allows for *fact-sensitive normative* theoretical inquiries without embracing idealised ideas about reality or some kind of descriptivist realism. It also defends the fact-sensitive approach to normativity as a fruitful strategy for the study of moral responsibility for climate change about which Parts 2 and 3 are concerned.

More particularly, the purpose of Part 1 is to develop a fact-sensitive normative account of moral responsibility. Because few scholars would deny the relevance of some facts in normative political thinking, the pertinent question is *how* facts are considered relevant and *which* types of facts are relevant. The meta-theoretical arguments for fact-sensitive normativity reveal a particular understanding of the relevance of facts in normative thinking. The main objective of this analysis is to demonstrate that it is possible for the fact-sensitive political theory to sustain a robust normative foundation whilst at the same time be responsive to factual circumstances.

The fact-sensitive political theory is developed through three arguments; a chapter is devoted to each. First, I show why it makes sense to distinguish abstract from concrete normative principles and argue in favour of a fact-sensitive approach to political theory. Second, I argue in Chapter 2 that the fact-sensitive approach allows for a restructure of the global justice debate; more particularly, I show that the abstract principles fail to determine who ought to do what ought to be done. Third,

I propose in Chapter 3 a fact-sensitive account of who ought to do what ought to be done which clarifies the premises of moral responsibility. Chapter 3 provides the bridge between the meta-theoretical discussion in Part 1 and the approach to moral responsibility in Parts 2 and 3.

In Part 2, I show that the examination of the fitness-conditions for moral responsibility for climate change is closely linked to (1) the understanding of what the political problems of climate change are and thus the suggested policy solutions and (2) the understanding of agency and rationality. The point of departure is that we cannot assume without further justification that the relevant morally responsible agents are individuals. It might as well be collective group agents, such as private corporations or democratic institutions, national states, or supra-national organisations. In order to establish a plausible relationship between agent and outcome, a theoretical account of agency, rationality, and explanatory theory is required. The reason to conduct this analysis is that there is no sociological consensus about rationality, agency, and causality. In order to justify ought-assignment with regard to moral responsibility for climate change, a solid theory of which type of agency is fit to be held responsible is necessary.

Thus, the account of fitness-conditions of responsibility takes its point of departure from a methodological debate about agency, rationality, and explanation. As Chapters 4 and 5 discuss, the methodology of *rational-individualism* perceives the current political environment of democratic institutions as a key obstacle to comprehensive climate action, whereas what I call the *non-reductionist institutionalist* methodology perceives the democratic decision-making process as a key solution to the current failures of climate action. Within the latter framework, the political challenge of climate change is not a 'wicked problem', but a politically manageable 'tame problem' which should be addressed with many overlapping 'clumsy solutions' derived from the entire toolbox of democratic governance (Rayner 2006, 10; Prins and Rayner 2006, v).[11] Conceived this way, climate change is not primarily *a tragedy of the commons* as is sometimes argued (Hardin 1968; Ostrom 1990; Gardiner 2011). In contrast, the political challenge of environmental harm and climate change is what I shall call a *tragedy of the few* (Ostrom et al. 1994; Schlager et al. 1994; Sagoff 2008; Shearman and Smith 2007; Desombre and Barkin 2011).

The inquiries of fitness for responsibility for climate change touch upon the current critique of the strength of democratic governance. Are democratic decision makers capable of developing policies and laws suitable for managing complex matters such as climate change? In his seminal article, Garrett Hardin (1968) argued that the collective problems of environmental and climate regulation cannot be solved by technical means only – they require novel political solutions. Today, however, the critique has turned around. Scholars, scientists, and politicians increasingly question the ability of democracy to mitigate climate change by political means and suggest technical and authoritarian solutions. From a situation in which climate activists once talked about the 'inconvenient truth', scientific experts have now become increasingly uneasy with the 'inconvenient democracy' (Stehr 2013; see also Hardin 1968; Ophuls 1977; Baber and Bartlett 2005;

Shearman and Smith 2007; Held and Hervey 2009; Beeson 2010; Lovelock 2010; Dryzek and Stevenson 2011; Persson and Savulescu 2012; Liao et al. 2012).

In Chapter 6, I discuss the critique of democratic governance. I argue that an important intellectual resource for understanding the critique of democratic climate governance is to be found in the current underestimation of the intermediating factors and the institutional capacity of contemporary democratic societies to handle climate challenges collectively and politically. Institutional capacity is an institution's capacity to facilitate, monitor, and implement policies in a legitimate, effective, and transparent manner.[12] The concept of institutional capacity rests on a non-reductionist institutionalist methodology (Shue 1980; Ostrom 2005; Le Grand 2006; Olsen 2007; Searle 2010; Hurley 2011; List and Pettit 2011). By recognising the non-reductionist methodology and the theory of group agency, new policy options can be offered.

Part 3 discusses the normative criteria for assigning moral responsibility to particular agents, specifically whether the agents can be morally blamed or excused. Whether we can justify the assignment of moral responsibility to groups is also discussed. In Chapter 7, I outline the normative criteria for the assignment of climate responsibility and stress the importance of looking at the agent's dynamic duties. These duties hold agents responsible for not only what agents ought to do but also whether they made the rights choices at an earlier point in time, which enables them to conduct what they ought to do now. I also elaborate on the different properties of individual, joint, and collective responsibility. By applying the normative criteria of moral responsibility, I argue in Chapter 8 that democratic citizens should be held *jointly* but not *individually* responsible for climate change, and in Chapter 9 that the so-called *occupants of institutionalised roles* should be held *collectively* responsible for the political failures of climate action. The discussion of collective responsibility also considers at what level of democratic climate governance ought to be implemented, and I discuss whether the occupants of institutionalised roles can also be held collectively responsible for not sustaining and improving the democratic quality of contemporary policy institutions.

Notes

1 The Montreal Protocol paved the way for banning the production of CFCs used in mundane applications, such as aerosol cans, refrigeration, and air conditioning equipment. The aim of the Kyoto Protocol was to stabilise the 'greenhouse gas concentrations in the atmosphere at a level that would prevent dangerous anthropogenic interference with the climate system' (UNFCCC 1992, Article 2).
2 CFCs' emission levels peaked in 1994. Since then, gases have been gradually removed: it is estimated that the Antarctic ozone layer will have recovered by 2050 or 2068 (Newman et al. 2006). The opposite applies in the case of the emission of GHGs, which is steadily growing. The global annual emission of around 40–50Gt of GHGs is emitted in large part by industry and power and transportation sectors (approximately 60 percent), and agriculture, land use, and deforestation (approximately 30 percent) (Nicholas Stern 2006, 1; 2010, 46; IPCC 2007, 36). GHG emissions are closely inspected due to their long-term and global effects (IPCC 2007, 39). GHGs induce changes in average weather conditions, global warming, and climate change.

3 The 'Conference of the Parties' (COP) is the governing body of the Convention on Biological Diversity and advances implementation of the Convention through the decisions it makes at its periodic meetings. The Convention was signed by 150 government leaders at the 1992 Rio Earth Summit and is dedicated to the promotion of sustainable development.

4 For a general overview of anthropocentric and non-anthropocentric versions of liberalism, see Hassoun (2011). Page distinguishes between anthropocentrism and three more detailed versions of eco-centrism – namely, zoo-centrism, bio-centrism, and eco-centrism (Page 2006, 139–141).

5 Sceptics argue that the changes of the planet's climate and atmosphere have positive consequences for new species and better living conditions for certain types of animals. 'To the conservation biologist, there is little positive to be said about extinction. From an evolutionary perspective, however, extinction is a double-edged sword. By definition, extinction terminates lineages and thus removes unique genetic variation and adaptation. But over geological time scales, it can reshape the evolutionary landscape in more creative ways, via the differential survivorship of lineages and the evolutionary opportunities afforded by the demise of dominant groups and the postextinction sorting of survivors' (Jablonski 2001, 5359). Furthermore, the possible beneficiaries of an average rise in temperature confined to two or three degrees include, for example, Russian, Canadian, and northern European farmers (de Perthuis 2011, 36).

6 There are other aspects of responsibility such as legal liability responsibility which in this analysis is set aside.

7 Synthesis studies show that among peer-reviewed papers on climate change, 97 percent of all studies exhibit a consensus that climate change has both social and anthropogenic causes (IPCC 2013). In today's geological science, it has become commonplace to talk about the current geologic time as *anthropocene* because the influence of human behaviour on the earth's atmosphere in recent centuries is so significant as to constitute a new geological epoch for its lithosphere (Zalasiewicz et al. 2010).

8 See, e.g., Williams (2005), Geuss (2008), Miller (2008) and Philp (2012). For recent contributions offering valuable overviews of the realistic debates and the ideal and non-ideal theory discussions, see Lægaard (2006), Ypi (2012), Valentini (2012), Hamlin and Stemplowska (2012), and Erman and Möller (2013).

9 Idealism is associated with Kant and neo-Kantianism, among others, Rawls (1971), Habermas (1984), and Cohen (2008).

10 For example, Machiavelli (2005), Hobbes (1991), Schmitt (1927), and Geuss (2008).

11 A 'tame problem' comprises 'complicated, but with defined and achievable end-states'. A 'wicked problem' comprises 'open, complex and imperfectly understood systems' which are difficult to manage politically (Prins and Rayner 2006, v).

12 The theory of institutional capacity refers to a literature on state capacity (Tilly 1992; Fukuyama 2012) and strong state-society relations (Migdal 1988).

References

Baber, W. F. and Bartlett, R. V., 2005. *Deliberative environmental politics: democracy and ecological rationality*. Cambridge, MA: MIT Press.

Barber, B. R., 2006. The politics of political science: 'value-free' theory and the Wolin-Strauss Dust-Up of 1963. *American Political Science Review*, 100 (4), 539–545.

Beeson, M., 2010. The coming of environmental authoritarianism. *Environmental Politics*, 19 (2), 276–294.

Brown, J., 1977. Moral theory and the ought-can principle. *Mind, New Series*, 86 (342), 206–223.

Bulkeley, H., 2013. *Cities and climate change*. New York, NY: Routledge.

Caney, S., 2010a. Climate change, human rights and moral thresholds. *In:* S. Humphreys, ed. *Human rights and climate change*. New York, NY: Cambridge University Press, 69–90. Reprinted in 2010: *In:* S. M. Gardiner, S. Caney, D. Jamieson, and H. Shue, eds. *Climate ethics: essential readings*. Oxford: Oxford University Press, 163–180.

Cohen, G. A., 2008. *Rescuing justice and equality*. Oxford: Oxford University Press.

Erman, E. and Möller, N., 2013. Three failed charges against ideal theory. *Social Theory & Practice*, 39 (1), 19–44.

Eshleman, A., 2009. Moral responsibility. *In:* E. N. Zalta, ed. *The Stanford encyclopedia of philosophy*. Winter edn. Palo Alto, CA: Stanford University. Available from: https://stanford.library.sydney.edu.au/archives/win2009/entries/moral-responsibility/ [Accessed 3 April 2019].

Fukuyama, F., 2012. *The origins of political order: from pre-human times to the French Revolution*. New York, NY: Farrar, Straus and Giroux.

Gardiner, S. M., 2010. Ethics and global climate change. *In:* S. M. Gardiner, S. Caney, D. Jamieson, and H. Shue, eds. *Climate ethics: essential readings*. Oxford: Oxford University Press, 3–38.

Gardiner, S. M., 2011. *A perfect moral storm: the ethical tragedy of climate change*. Oxford: Oxford University Press.

Geuss, R., 2008. *Philosophy and real politics*. Princeton, NJ: Princeton University Press.

Habermas, J., 1984. (English translation) *The theory of communicative action*. Boston, MA: Beacon Press. *In:* German: Habermas, J., 1981. *Theorie des kommunikativen Handelns*. Frankfurt am Main: Suhrkampf.

Hamlin, A. and Stemplowska, Z., 2012. Theory, ideal theory and the theory of ideals. *Political Studies Review*, 10 (1), 48–62.

Hardin, G., 1968. The tragedy of the commons. *Science*, 162 (3859), 1243–1248.

Hassoun, N., 2011. The anthropocentric advantage? Environmental ethics and climate change policy. *Critical Review of International Social and Political Philosophy*, 14 (2), 235–257.

Held, D. and Hervey, A. F., 2009. *Democracy, climate change and global governance democratic agency and the policy menu ahead*. Policy Network Paper, November. London: Policy Network.

Hobbes, T., 1991/1651. *Leviathan*. R. Tuck, ed. Cambridge: Cambridge University Press.

Hoff, J. and Strobel, B. W., 2013. A municipal 'climate revolution'? The shaping of municipal climate change policies. *The Journal of Transdisciplinary Environmental Studies*, 12 (1), 3–14.

Hurley, S., 2011. The public ecology of responsibility. *In:* C. Knight and Z. Stemplowska, eds. *Responsibility and distributive justice*. Oxford: Oxford University Press, 187–215.

ICLEI Global Reports, 2010. *Cities in a post-2012 climate policy framework*. Climate Financing for City Development? Views from Local Governments, Experts and Businesses. Available from: www.iclei.org/fileadmin/PUBLICATIONS/Papers/Cities_in_a_Post-2012_Policy_FrameworkClimate_Financing_for_City_Development_ICLEI_2010.pdf [Accessed 15 January 2017].

IEA, 2011. *World energy outlook 2011*. Paris: OECD/IEA.

IPCC (Intergovernmental Panel on Climate Change), 2007. *Climate change 2007: synthesis report: contribution of Working Group I, II and III to the Fourth Assessment Report of the International Panel on Climate Change*. R. K. Pachauri and A. Reisinger, eds. Geneva: IPCC, 25–73. Available from: www.ipcc.ch/publications_and_data/ar4/syr/en/contents.html [Accessed 20 January 2017].

IPCC (Intergovernmental Panel on Climate Change), 2013. *Summary for policymakers*. Twelfth Session of Working Group I Approved Summary for Policymakers. Available from: www.climatechange2013.org/images/uploads/WGIAR5-SPM_Approved-27Sep2013.pdf [Accessed 20 January 2017].

Jablonski, D., 2001. Lessons from the past: evolutionary impacts of mass extinctions. *PNAS*, 98 (10), 5393–5398.

Knight, C. and Stemplowska, Z., eds., 2011. *Responsibility and distributive justice*. Oxford: Oxford University Press.

Lægaard, S., 2006. Feasibility and stability in normative political philosophy: the case of liberal nationalism. *Ethical Theory and Moral Practice*, 9 (4), 399–416.

Le Grand, J., 2006. *Motivation, agency, and public policy: of knights and knaves, pawns and queens*. New York, NY: Oxford University Press.

Liao, S. M., Sandberg, A., and Roache, R., 2012. Human engineering and climate change. *Ethics, Policy and Environment*, 15 (2), 206–221.

List, C. and Pettit, P., 2011. *Group agency: the possibility, design, and status of corporate agents*. Oxford: Oxford University Press.

Lovelock, J., 2010. James Lovelock on the value of sceptics and why Copenhagen was doomed. Interviewed by Leo Hickman. *The Guardian*, Monday, 29 March. Available from: www.theguardian.com/environment/blog/2010/ mar/29/james-lovelock [Accessed 23 September 2013].

Machiavelli, N., 2005/1532. *The Prince with related documents*. 1st edn. W. J. Cornell, ed. Boston, MA: Bedford St. Martins.

Migdal, J., 1988. *Strong societies and weak states: state-society relations and state capabilities in the third world*. Princeton, NJ: Princeton University Press.

Miller, D., 2008. Political philosophy for earthlings. *In:* D. Leopold and M. Stears, eds. *Political theory: methods and approaches*. Oxford: Oxford University Press, 29–48.

Miller, S., 2010. *The moral foundations of social institutions: a philosophical study*. Cambridge: Cambridge University Press.

Nagel, T., 1995. Personal rights and public space. *Philosophy and Public Affairs*, 24 (2), 83–107.

Newman, P. A., Nash, E. R., Kawa, S. R., Montzka, S. A., and Schauffler, S. M., 2006. When will the Antarctic ozone hole recover? *Geophysical Research Letters*, 33 (12), L12814.

Olsen, J., 2007. *Europe in search of political order: an institutional perspective on unity/diversity, citizens/their helpers, democratic design/historical drift and the co-existence of orders*. Oxford: Oxford University Press.

O'Neill, O., 1988. Ethical reasoning and ideological pluralism. *Ethics*, 98 (4), 705–722.

O'Neill, O., 2013. *Acting on principle: an essay on Kantian ethics*. Cambridge: Cambridge University Press.

Ophuls, W., 1977. *Ecology and the politics of scarcity*. San Francisco, CA: Freeman.

Ostrom, E., 1990. *Governing the commons: the evolution of institutions for collective action*. Cambridge: Cambridge University Press.

Ostrom, E., 2005. *Understanding institutional diversity*. Princeton, NJ: Princeton University Press.

Ostrom, E., Gardner, R., and Walker, J., 1994. *Rules, games, and common-pool resources*. Ann Arbor, MI: University of Michigan Press.

Page, E., 2006. *Climate change, justice and future generations*. Cheltenham: Edward Elgar.

Persson, I. and Savulescu, J., 2012. *Unfit for the future: the need for moral enhancement*. Oxford: Oxford University Press.

Philp, M., 2012. Realism without illusions. *Political Theory*, 40 (5), 629–649.

Plant, R., 2009. *The neoliberal state*. Oxford: Oxford University Press.

Prins, G. and Caine, M. E., 2013. *The vital spark. Innovating clean and affordable energy for all*. The Third Hartwell Paper, July. London: LSE Academic Publishing.

Prins, G. and Raynor, S., 2006. *The wrong trousers: radically rethinking climate policy*. Available from: eureka.sbs.ox.ac.uk/66/ [Accessed 26 August 2017].

Rawls, J., 1971. *A theory of justice*. Cambridge, MA: The Belknap Press of Harvard University Press.

Rawls, J., 1999. *The law of peoples: with the idea of public reason revisited*. Cambridge, MA: Harvard University Press.

Rayner, S., 2006. *Wicked problems: clumsy solutions*. First Jack Beale Memorial Lecture, University of New South Wales, Sydney, Australia, 25 July 2006.

Rose, G. L., 2011. *Gaps in the implementation of environmental law at the national, regional and global level*. First Preparatory Meeting of the World Congress on Justice, Governance and Law for Environmental Sustainability, UNEP, 1–30.

Sagoff, M., 2008. *The economy of the earth, philosophy, law and the environment*. New York, NY: Cambridge University Press.

Schlager, E., Blomquist, W., and Yan Tang, S., 1994. Mobile flows, storage, and self-organized Institutions for governing common-pool resources. *Land Economics*, 70 (3), 294–317.

Schmitt, C., 1927/2006. *The concept of the political*. Chicago, IL: University of Chicago Press.

Searle, J., 2010. *Making the social world: the structure of human civilization*. New York, NY: Oxford University Press.

Sen, A., 2009. *The idea of justice*. London: Penguin Books.

Shearman, D. and Smith, J. W., 2007. *The climate change challenge and the failure of democracy*. Westport, CT: Praeger.

Shue, H., 1980. *Basic rights: Subsistence, affluence, and U.S. foreign policy*. 2nd edn. Princeton, NJ: Princeton University Press.

Shue, H., 1999. Global environmental and international inequality. *International Affairs*, 75 (3), 531–545. Reprinted in 2010: *In:* S. M. Gardiner, S. Caney, D. Jamieson, and H. Shue, eds. *Climate ethics: essential readings*. Oxford: Oxford University Press, 101–199.

Singer, P., 1993. *Practical ethics*. Cambridge: Cambridge University Press.

Singer, P., 2010. One atmosphere. *In:* S. M. Gardiner, S. Caney, D. Jamieson, and H. Shue, eds. *Climate ethics: essential readings*. Oxford: Oxford University Press, 181–199.

Stern, N., 2006. What is the economics of climate change? *World Economics*, 7 (2), 1–10.

Stehr, N., 2008. Introduction: is freedom a daughter of knowledge? *In:* N. Stehr, ed. *Knowledge and democracy: a 21st century perspective*. New Brunswick, NJ: Transaction Publishers, 1–8.

Stern, N., 2010. The economics of climate change. *In:* S. M. Gardiner, S. Caney, D. Jamieson, and H. Shue, eds. *Climate ethics: essential readings*. Oxford: Oxford University Press, 39–76.

Stehr, N., 2013. An inconvenient democracy: knowledge and climate change. *Society*, 50, 55–60.

Streeck, W., 2011. The crises of democratic capitalism. *New Left Review*, 71, 5–29.

Tilly, C., 1992. *Coercion, capital and European states, ad 990–1992*. Cambridge, MA: Blackwell.

UNFCCC, 1992. *United Nations Framework Convention on Climate Change*. Available from: http://unfccc.int/resource/docs/convkp/conveng.pdf [Accessed 6 March 2018].

Valentini, L., 2012. Ideal vs. nonideal theory: a conceptual map. *Philosophy Compass*, 7 (9), 654–664.

Williams, B., 2005. *In the beginning was the deed: realism and moralism in political argument*. Princeton, NJ: Princeton University Press.

WWF. 2012. *Living Planet Report 2012 Summary*. Available from http://awsassets.panda.org/downloads/lpr_2012_summary_booklet_final.pdf [Accessed 15 June 2019].

Ypi, L., 2012. Facts, principles and the third man. *Socialist Studies*, 8 (1), 198–215.

Zalasiewicz, J., Williams, M., Steffen, W., and Crutzen, P., 2010. the new world of the anthropocene. *Environment Science & Technology*, 44 (7), 2228–2231.

Part 1

Fact-sensitivity and normativity

1 Fact-sensitive political theory

The goal of Part 1 is to show that the fact-sensitive account of normativity allows for an integration of facts into the normative theory whilst sustaining a robust justification in fundamental principles. More particularly, the fact-sensitive account is a fruitful approach to climate responsibility because concrete ought-assignments require a fact-sensitive component.

The fact-sensitive account is developed over three chapters. In this first chapter, I elaborate on the distinction between abstract and concrete fact-sensitive principles by arguing that one of the main differences between these two principles concerns the contingent on abstract or concrete facts. In Chapter 2, I demonstrate that abstract principles fail to determine *who* ought to do *what*, and in Chapter 3, I elaborate on the link between the fact-sensitive normativity (Part 1) and moral responsibility for climate change (Parts 2 and 3) by defending how the fact-sensitive normativity provides a suitable framework for identifying the normative criteria for *can-conditions* for fulfilment of ought-statements. In Part 3, I argue that this has implications for how to construct the normative theory of moral responsibility for climate change.

Concrete and abstract principles

The purpose of the following section is to show how fact-sensitive normativity can be understood as concrete principles which are contrasted to abstract principles but nonetheless dependent on the latter. The difference between abstract and concrete principles is a question of the level of abstraction. According to Onora O'Neill's definition, abstraction is

> a matter of selective omission, of leaving out some predicates from descriptions and theories. Selective omission can hardly be objected to. It is unavoidable. No use of language can be fully determinate. Abstraction is a precondition for logic, for scientific reasoning, and for many highly respected (and lucrative) forms of practical reasoning, such as legal and commercial reasoning. Abstraction is also needed if we are to reason in ways that can be taken seriously by others who disagree with us. By abstracting we may succeed in reasoning in ways that are detachable from commitment to the

full detail of our own beliefs. The less abstract our reasoning, the greater the likelihood that it hinges on premises that others will dispute and that its conclusions will seem irrelevant to those others.

(O'Neill 1988, 11)

When we embrace this notion of abstraction, which I believe we should, we can distinguish between abstract and concrete principles dependent on the level of abstraction. An example of the former could be that *each human being is entitled human rights*, whilst an example of the latter could be that *global institutions should be democratic*. A different way to put it is to say that the abstract moral principles aim to define *what is good and bad*, whereas the concrete normative principles aim to define *what is just*.

Abstract normative principles are contingent upon abstract facts, whilst concrete principles are contingent upon concrete facts. In this context, I take abstract facts to be general facts, such as *there exist human beings* and *there is life on the earth*, with which most people would agree. In contrast, concrete facts are facts such as *nation-states are important political entities* and *global wealth is distributed unequally*, about which there is less agreement. The reason for this is that the relative importance of concrete facts about politics and the economy may not always apply.[1]

Concrete facts can be subdivided into two categories. One concrete set of facts concerns facts about political circumstances. This includes a broad set of concrete facts about nationality, stateness, citizenship, and democratic institutions. Another set of facts concerns facts about the feasibility and *can*-conditions of agency. In contrast to the concrete principles that are contingent upon concrete facts about the world and political systems, the principles primarily concerned with the normative criteria for sound ought-assignment to particular agents are concrete ethico-normative ought-judgements.

An important premise for concrete ought-judgements is methodologically embedded theories about human agency and rationality. These ought-judgements can be contrasted with the abstract ought principles which say something about the general principle what ought to be done. A different way to stress the difference is to say that the purpose of the abstract ought-judgements is to answer the question of *what ought to be done*, whilst the purpose of the concrete ought-judgements is to answer the question of *who ought to do what*. By combining the two distinctions, it becomes possible to differentiate four types of normativity, based on two distinctions between (1) abstract and concrete principles and (2) moral principles and ought-judgements (cf. Table 1.1).

Table 1.1 Four types of normativity

	Moral principles	*Ought-judgements*
Abstract facts	1. Abstract fundamental principles	3. Abstract ought principles
Concrete facts and agency assumptions	2. Concrete normative principles	4. Fact-sensitive ought-assignments

In Chapter 3, I discuss the concrete *ethico-normative principles* in contrast to the abstract ones. It should be noted that one reason for why it is essential to outline the four types of normativity is that abstract moral principles are frequently confused with concrete ought-assignments. This is unfortunate because the former depends on highly abstract ideas about humanity, whilst the latter depend on methodologically embedded ideas about human agency under specific circumstances. This distinction has great implications for the discussion of who ought to be held morally responsible for climate change.

Facts and principles

Two related points should be considered with regard to the definition of concrete fact-sensitive principles. The first is that these normative principles are dependent on concrete facts; the second is that they are dependent on abstract principles. Let us consider the dependency between concrete fact-sensitive principles and facts first.

One strategy is to argue that the common distinction between *descriptive judgements* and *normative judgements* has only analytical purposes (Putnam 2002, 13). Whilst it might be possible to distinguish between the two, what the fact-sensitive political theory is interested in are the relevant research questions that lie in what some have called, 'the tension field between *is* and *ought*' (Rasmussen 1971, 47; Rasmussen 1989, 461). Consider one example. Climate change is an important research area for social and political theory because it is factually important and a hard case with regard to its normative implications. Put differently, the factual and normative aspects of climate change depend on each other: we are morally concerned about climate change because of the factual changes in the earth's climate and atmosphere, and we are factually concerned about climate change because we have several moral principles that suggest climate change is morally wrong.

One way to provide some plausibility for what we could call 'the dependency thesis' is to accept that some types of judgements about facts and norms cannot be completely disentangled. In this regard, Hilary Putnam offered a productive theorisation for this. In order to understand Putnam's argument, I find it fruitful to point to two aspects: the study of *that p* is dependent on epistemic values for *what is good research*. Scientific knowledge presupposes epistemic values such as plausibility and cohesion (Putnam 2002, 13). Putnam then argued more controversially that the study of *that p* depends on normative values about *what is a good society*.

Putnam argued in support of his second point that scientific descriptions of facts frequently use words and concepts whose content is not merely descriptive but contains evaluation aspects as well. He calls these types of words and concepts 'ethical concepts' (Putnam 2002, 34–37). According to Putnam, a description of a political dictator as 'evil' is not only an empirical description but also contains a moral evaluation of whether the acts and policies of the dictator are morally unjustifiable (Putnam 2002, 37). More generally, he argued that knowledge about empirical facts presupposes 'knowledge about values' (Putnam 2002, 136–137). However, when we consider the dictator example, it might be more fruitful to suggest that we assume a symmetrical relationship between this specific set of facts and values in the sense that the facts about the dictatorship presuppose values and values against dictatorships presuppose facts.

An opponent of the dependency thesis, however, might argue that it is possible to study the number of neo-Nazis in Denmark, for example, independently of whether it is presumed that it is good or bad that there are few or many Nazis. This argument implies that empirical judgements (*that p*) can be justified and studied independently of whether *it is good or bad that p* because the values are 'irrelevant for the answering of empirically relevant questions' (Lippert-Rasmussen 2005a, 284). Proponents of the dependency thesis, however, may re-join this claim: they might ask whether it makes sense to talk about what is 'empirically relevant' independently of what may be politically and morally relevant.

How do we determine what is empirically relevant if no political or moral standards are allowed? Is it possible to claim that the question of empirical relevance depends on the question of what is politically wishful, legitimate, morally permissible, and acceptable? If this is accepted, it is unclear whether there is independency between empirical relevance and normative relevance. Thus, the proponent of the dependency thesis can maintain that the empirical study of Nazis in Denmark contains a judgement about whether it is good or bad that there are many or few Nazis and moreover that it is the normative judgement that makes the empirical study relevant to pursue. The argument here is that the empirical relevance is dependent on the political and moral significance of the study.

But is the Nazi example a strong argument that empirical judgements cannot be kept distinct from normative judgements? If we talk about attributes such as evil, good, and the like, it seems plausible that *is* and *ought* may overlap in ethical concepts. It becomes more difficult if the description does not concern a dictator, but concerns complex cases such as climate change. How do we know whether climate change is bad or good? Or even more difficult, how do we know whether certain acts and policies that might (indirectly) contribute to climate change are good or bad? In these cases, it might be impossible to identify an overlap between the descriptive and evaluative components of the concept. Do we know that certain acts in themselves are good or bad? Does the concept of climate unfriendliness contain ideas about whether it is a politically and morally good or bad property?

Relational value judgements

A different way of framing this challenge is to say that the concept does not in itself contain these judgements but requires *relational* value judgements (Lippert-Rasmussen 2005a). Here we come to the second question of how concrete fact-sensitive norms can be justified in abstract principles.

Relational value judgements can be conceptualised in different ways. We can distinguish a *non-fundamental* from a *fundamental* type of value judgements. The non-fundamental type is commonly known as coherentism, which takes all justification to be inferential, which means that all judgements cohere with other beliefs. Yet consider this difficulty. How to justify value A if A's validity depends on other values B and C? If A justifies B and C, and B and C justify A, the justification is circular. If A justifies B and B justifies C, etc., the challenge might be an infinite regress (Lippert-Rasmussen 2005a, 276). On the other hand, the fundamental variant takes fundamental judgements to be non-inferentially justified – that is, they are not derived from other value judgements.

However, there might be a third option, a *hybrid* position which takes fundamental judgements to be non-inferentially justified but at the same time appreciates that non-fundamental value judgements are justified by virtue of their relational value and coherence with the fundamental value judgements. The hybrid position cannot be used as an argument for the distinction between abstract and concrete normative principles since the distinction is presumed in the definition. Nonetheless, it provides reasons for why it may be a plausible and fruitful approach to political theoretical considerations in which concrete fact-dependent judgements can be justified in fundamental principles. By presuming a set of abstract fundamental principles, the concrete normative judgements about, for example, the proposition *politics is climate unfriendly* allow for a fundamental justification. The present task is to show that the concrete normative judgements have the potential to make normatively robust evaluations of factual matters.

Let us consider the climate-friendliness example again. Let us presume a set of fundamental principles about human rights and then assert the fact that the politics is climate unfriendly. Now the question is whether the fundamental principle of human rights can be used in the normative evaluation of whether it is good or bad that the politics is climate unfriendly. One pertinent question is whether we have reason to believe that the climate unfriendliness is (in)compatible with the protection of human rights. In order to answer this question, we need a further normative principle, which may be the following concrete normative judgement: *if basic human rights are appreciated, it is bad that politics is climate unfriendly because it may undermine the natural and human conditions for upholding democratic and law institutions, which are important for the protection of basic human rights.* This fact-sensitive judgement is a normative judgement that is fact-dependent but at the same time warranted by fundamental moral principles.[2]

Within this understanding, sound normative arguments are provided in light of existing institutional, social, and political facts. The argument is expressed in the following relational form: *if Q, it is good or bad that p*, where Q is an p-independent abstract normative principle and p a concrete fact. This means that the fundamental principle (Q) is independent of the concrete facts (p) which are relevant for one particular concrete fact-sensitive principle.

This account of fact-sensitivity presumes a weak reading of G. A. Cohen's argument for fact-insensitivity. Cohen's concept of fact-insensitive principles provides important theoretical resources that help us comprehend how fundamental principles can justify concrete fact-sensitive principles. He argued in particular that fact-sensitive principles can only 'reflect or respond to a fact' because they presume 'a principle that is not a response to a fact' (Cohen 2008, 230) and that 'there is always an explanation why any ground grounds what it grounds' (ibid.,

236). Cohen has been criticised for suggesting an essentialist and idealist account of justice (Nielsen 2012, 218). Cohen, however, is clear that he remains neutral as to which theorisation of fundamental principle is the right one (Cohen 2008, 257). A more serious critique of Cohen's argument is that the fact-insensitive thesis fails due to an infinite regress (Ypi 2012). This critique assumes that fundamental normative principles are completely fact-independent, which I call the 'strong view'.

In contrast, I suggest a 'weak reconstruction' of Cohen's distinction between fact-insensitive and fact-sensitive principles by redefining the distinction as being one of abstract and concrete principles. The weak reconstruction suggests that fact-insensitive principles are abstract principles that are only independent of a concrete set of facts, but not abstract sets of facts. This argument makes it possible to warrant fact-sensitive principles in fundamental normative principles without assuming a strong account of fact-insensitivity. Some passages of Cohen's discussion of the issue indicate that he could support the weak reconstruction. In one example, he says that '[a]lthough a principle that makes a fact matter, in the indicated fashion, is insensitive to whether or not that fact obtains, it may yet be sensitive to (other) facts' (Cohen 2008, 234). Notwithstanding Cohen's account, I conclude that the weak reconstruction provides the grounds to justify and warrant concrete fact-sensitive principles as relational value judgements in fundamental principles despite the fact that they are contingent on concrete facts.

Fact-sensitive political theory

One point should be added: the concrete fact-sensitive account of normativity, which claims that the fact-sensitive norms are dependent on facts and abstract fundamental norms, remains neutral to the metaphysical status of abstract fundamental normative principles. The reason for this position is that concrete normative judgements are presumed to be one derived from political theory, not from moral philosophy (ibid., 263), a claim that presumes that there is a relevant distinction between political theory and moral philosophy to be made. Thus, two definitions can be distinguished: one that is occupied with the definition, nature, and validity of moral principles, and one that is occupied with how principles engage with facts.

Because there is still no agreement about the metaphysical status of the abstract normative principles, it is fruitful to distinguish those inquiries from the inquiries of how fact-sensitive principles response to facts. This is because the argument for concrete principles presumes the existence of abstract principles. If abstract principles are not presumed, it is impossible to engage in fact-sensitive normative examinations. As Thomas Nagel argued, 'Not everything can be revised [at once], because something must be used to determine whether a revision is warranted – even if the proposition at issue is a very fundamental one' (Nagel 1995, 65; cited from Cohen 2008, 243n18).

Thus, if fact-sensitive normative inquiries are not appreciated as a sub-discipline of political philosophy, they will be squeezed by on-going debates on the ontological status of abstract moral principles, about which there remains considerable disagreement. In short, these fact-sensitive examinations of norms are difficult to pursue if one does not allow for distinctions between abstract and concrete norms. These rules may be justified by fundamental principles, but these abstract considerations of justice do not exhaust the normative evaluation of concrete rules of regulation.

Andrew Mason (2004) made a similar point. He argued that it is fruitful to distinguish a number of different levels of analysis of ideals, politics, and justice. One level is abstract; another level takes feasibility and balancing concerns into consideration. The important lesson here is that the two levels complement each other and should not be collapsed because the two are conceptually and normatively different (Mason 2004, 265–266). By distinguishing between two aspects of normativity, the examination of the ontological status of principles of justice can be bracketed from the normativity of fact-dependent acts and political practices without neglecting the importance of the former.

Thus, by assuming the existence of fundamental normative principles, *fact-sensitive political theory* establishes a venue for the discussion of concrete normative questions: what is politically relevant, what should be done, and what acts should be promoted? However, since these questions cannot be answered at the level of concrete normativity alone, the important question is which fact-sensitive normative principles can be warranted in fundamental moral principles. Cohen argued that in order to comprehend fact-sensitive normativity, 'we necessarily have to recourse to basic principles to justify the rules of regulation that we adopt' (Cohen 2008, 267). Seen in this light, abstract normative principles are lenses through which the interface between facts and concrete normative principles can be illuminated. The idea is that we become aware of a fact's normative implications in light of abstract principles, and we become aware of concrete principles when studying fundamental principles in light of facts.

One might still be unsatisfied with the fact that concrete fact-sensitive principles are not validated in a pure normative way. People may disagree about which concrete norms cohere with abstract norms. How can we then be sure that the defended concrete normative principles are the right ones? The short answer is that we cannot be completely sure. One further implication of this is that normative inquiries about factual matters constitute an on-going process in which wrong or incomplete concrete normative interpretations should be re-joined or advanced. Some may use these challenges of justification as an argument that neither concrete nor abstract are relevant topics to investigate. Rather, we should conclude the opposite. Because we may disagree about concrete principles and how they may be justified in abstract principles, we should keep scrutinising concrete principles and examining how they may be justified.

I conclude that the fact-sensitive political theory presents a coherent model for investigating normativity in light of facts, politics, and social practices without turning to idealism or realism. We need to avoid idealism because concrete facts are considered necessary; realism and relativism are avoided by the valid normative warrant of the fact-sensitive principles in the abstract norms.

Disagreements about facts

One further implication of the fact-sensitive account of normativity is that a strict set of facts about what politics is and how agents are behaving cannot be presumed. This is opposite to the position of proponents of political realism who take a strict account of facts. The rejection of a strict set of facts is particularly pertinent with regard to ought-assignments, where the relevant facts are methodologically embedded ideas of agency, rationality, and causality. Put differently, the fact-sensitive approach to normative challenges argues that facts about human agency are not strict facts: they are normatively embedded notions on ought-assignments that depend on the factual circumstances and methodological assumptions by which human agency is conceived.

However, misunderstandings at the lower levels of fact-sensitivity are likely to occur because people may address different sets of facts and favour different sets of political norms. This shatters the soundness of fact-sensitive principles. Thus, in order to avoid unjustified interpretations, it remains essential, as noted earlier, to warrant the concrete principles in normative principles of higher levels of abstraction that are independent of the concrete facts upon which the concrete principle relies (Cohen 2008, 234). We cannot decide on concrete normative principles without assuming a set of abstract normative principles that warrant the concrete normative interpretation of the factual state of affairs (Cohen 2008, 230).

One example of this concern is reflected in Cohen's critique of Rawls's theory of distributive justice and the difference principle (Cohen 1997; Cohen 2008).[3] Rawls's point is that fundamental principles of justice are responsive to the facts of the human condition: 'There is no objection to resting the choice of first principles upon the general facts or economics and psychology'. Furthermore, the difference principle 'relies on the idea that in a competitive economy (with or without private ownership) with an open class system excessive inequalities will not be in the rule' (Cohen 2008, 259; see also Rawls 1971).

When Rawls's principles of justice are interpreted in the terminology of abstract and concrete principles, they may be regarded as concrete normative principles and the normative standards for evaluating low or excessive inequality as abstract normative principles. The question is whether Rawls understands the principles correctly; if not, he risks conflating the concrete fact-sensitive principles with the abstract fact-sensitive principles. To quote Cohen,

> It is a fundamental error of *A Theory of Justice* that it identifies the first principles of justice with the principles that we should adopt to regulate society. Rawls rightly says that 'the correct regulative principle for anything depends on the nature of that thing': facts are of course indispensable to the justification of rules of regulation. But rules of regulation necessarily lack ultimacy: *they* cannot tell us how to evaluate the effects by reference to which they themselves are to be evaluated. Sociology tells us what the effects of various candidate rules would be, but a normative philosophy that lacks sociological

input is needed to evaluate those effects and thereby to determine, jointly with sociology, what rules we should adopt.

(Cohen 2008, 265–266, emphasis in the original)

In this passage, Cohen elaborates Rawls's objective in order to identify a just societal basic structure that is compatible with human nature and the economic circumstances of modern societies. To obtain a just society, Rawls requires that people submit willingly to the standard of justice – i.e., the difference principle. He assumes an 'ethos of justice' that informs individuals that inequality is unjust (Cohen 2008, 16). Ostensibly, this reflects the distinction between the two different types of norms, where some are abstract whilst others are concrete norms. However, according to Cohen, Rawls misidentifies the abstract normative question *what is justice?* with the concrete normative question *what principles should be adopted to regulate our affairs?* (Cohen 2008, 269).[4] Cohen sums up the challenge in the following way:

> [F]acts undoubtedly help to decide what rules of regulation should be adopted, that is, legislated and implemented, if only because facts constrain possibilities of implementation and determine defensible trade-offs (*at the level of implementation*) among competing principles. But the principles that explain, with the facts, why a given set of principles is the right one to adopt don't reflect facts, and non-exposure of those more ultimate principles means failure to explain why we should adopt the principles that we should adopt.
>
> (Cohen 2008, 244, emphasis original)

According to the fact-sensitive account, Rawls wrongly argued that facts of incentives and competition have implications for justice (e.g., fundamental abstract principles) instead of concluding something about concrete fact-sensitive principles that relate to how to regulate our social behaviour and policy. This is unsatisfactory since the point of distinguishing abstract principles from concrete ones is that it is possible to draw attention to economic incentives and other factual circumstances of politics and agency without saying anything about the fundamental set of moral principles.

If proponents of liberalism neglect to differentiate abstract principles of freedom and equality from the concrete norms for individual incentives, they risk promoting principles that perceive personal incentives as aspects of fundamental principles of justice instead of as concrete feasibility and can-constraints that require a more fundamental justification before they can be used as justified constraints on moral requirements. In contrast, the fact-sensitive account of normativity takes factual constraints, such as stability, respect for economic equilibrium, and publicity, to be concrete normative notions that are in need of fundamental justification (Cohen 2008, 285).

In the next chapter, I illustrate this notion of the confusion of abstract and concrete principles further by drawing attention to the discussion in the contemporary literature on global justice. Here, we see several attempts to integrate factual matters into the construction of theoretical and normative models for global justice arguments. Many of these attempts, however, can be criticised for providing a wrong or incomplete link between norms and facts, which importantly frames the ethico-discussion of who ought to be held morally responsible for climate change.

Conclusion

Because concrete fact-sensitive principles are at the core of the fact-sensitive account of normativity, the idea that the normative and factual theorisations in political theoretical dilemmas should not be completely disentangled. In the following chapter, I turn to the indeterminacy challenge of the abstract normative principles. In Chapter 3, I argue that concrete principles are capable of providing an account of who ought-assignment should be attributed to, which frames the investigations of fitness-conditions and moral responsibility for climate change in Part 2 and the normative criteria in Part 3.

Notes

1 Note that we do not need to assume that there is a clear demarcation between general and concrete facts. There is a grey area of facts between the clearly abstract and clearly non-abstract facts that shift subgroup depending on what question is at stake and what level of abstraction is proposed.
2 If we grant that this argument is on the right track, an important concrete and fact-sensitive question then becomes whether agent A in a concrete situation *should* select action or politics Z or X (Brecht 1959, 269). I return to this discussion in Chapter 3, which concerns the question of how to derive ought-statements from p-independent norms. I argue that it seems plausible that the concrete level of normative theory allows for judgements of the following kind: if we value political freedom as a fundamental principle (Q), the politicians (A) *should* not be corrupt, and they *should* choose politics Z (non-corruption) over politics X (corruption).
3 In this section, I set aside discussions of whether Cohen reads Rawls correctly. The goal here is to obtain a better understanding of Cohen's understanding of facts and norms, not of Rawls's theory of justice.
4 Following this line of reasoning, Thomas Pogge has suggested that Rawls should have called his book *A Theory of Regulation* instead of *A Theory of Justice* (Pogge 2008, 456). Note, however, that Rawls in the books *Laws of Peoples* (1999) and *Justice as Fairness* (2003) re-joins several of points that he made in *A Theory of Justice*. His notions on *fairness* and *non-ideal utopia* in particular resemble what in the present context is called concrete fact-sensitive principles.

References

Brecht, A., 1959. *Political theory: the foundations of twentieth-century political thought.* Princeton, NJ: Princeton University Press.
Cohen, G. A., 1997. Where the action is: on the site of distributive justice. *Philosophy and Public Affairs*, 26 (1) (Winter), 3–20.
Cohen, G. A., 2008. *Rescuing justice and equality.* Oxford: Oxford University Press.

Lippert-Rasmussen, K., 2005a. [Erik Rassmussen's relativism of worth: content and validity.] Erik Rasmussens værdirelativisme: indhold og gyldighed. *Politica*, 37 (4), 274–286.

Mason, A., 2004. Just constraints. *British Journal of Political Science*, 34 (2), 251–268.

Nagel, T., 1995. Personal rights and public space. *Philosophy and Public Affairs*, 24 (2), 83–107.

Nielsen, K., 2012. Rescuing political theory from fact-insensitivity. *Socialist Studies*, 8 (1), 216–245.

O'Neill, O., 1988. Ethical reasoning and ideological pluralism. *Ethics*, 98 (4), 705–722.

Pogge, T., 2008. Cohen to the rescue! *Ratio*, 21 (4), 454–475.

Putnam, H., 2002. *The collapse of the fact/value dichotomy and other essays*. Cambridge, MA: Harvard University Press.

Rasmussen, E., 1971. [Comparative Politics Vol. 1.] *Komperativ politik vol 1*. København: Gyldendal.

Rasmussen, E., 1989. [Relativism of worth – what does it mean? Commentary to a noncognitive opinion.] Hvad betyder 'værdirelativisme'? Kommentarer til et nonkognitivistisk debatindlæg. *Politica*, 21 (4), 456–462.

Rawls, J., 1971. *A theory of justice*. Cambridge, MA: The Belknap Press of Harvard University Press.

Ypi, L., 2012. Facts, principles and the third man. *Socialist Studies*, 8 (1), 198–215.

2   The indeterminacy challenge

Fact-sensitive accounts of moral responsibility for climate change fall within the theory of moral cosmopolitanism and global justice, more particularly, global environmental justice that relates to the global distribution of environmental burdens and benefits (Caney 2005b, 748). The main purpose of this chapter is to demonstrate that abstract cosmopolitan principles fail to determine concrete ethico-normative principles for *who ought to what*. First, I argue why abstract versions of moral cosmopolitanism fail what I call the *indeterminacy challenge*. Second, I discuss various models that might overcome the indeterminacy challenge. I conclude that that the concrete normative questions of *who ought to do what* and *at what political institutional level* depends on which meta-theoretical and methodological assumptions are assumed, which I elaborate on in Chapter 3.

Moral cosmopolitanism

Let us begin by outlining one standard definition of cosmopolitanism. Cosmopolitans argue that humans are 'world citizens' devoted to 'the moral community made up by the humanity of all human beings' (Nussbaum 1996, 7). The moral concepts important for cosmopolitanism are individualism and moral universalism. Individualism is the notion that the ultimate unit of moral concern is every human being. Such moral concern can be expressed as subjective goods such as human happiness, desire fulfilment, and preference satisfaction, or as objective goods, such as human need fulfilment, capabilities, resources, and opportunities (Pogge 1992, 49).

Moral universalism is defined by three underlying assumptions: (a) valid moral principles exist, (b) these moral principles apply to all persons who share some common morally relevant properties, and (c) persons throughout the world share some morally relevant similarities (Caney 2005a, 35–36). When we appreciate moral universalism, the moral value of various good and bad things that can happen to people should be valued in the same way, no matter who they are or where in the world they live. In other words, a world in which there is a starving person in Africa is as bad as a world in which there is a starving person in the US (D. Miller 2007, 28).

The ethico-normative implications of moral universalism and individualism are normally considered to be (1) 'that human beings are all subject to the same set of moral laws: we must treat others in accordance with those laws no matter where in the universe they live; they likewise must treat us in the same way' (D. Miller 2007, 24) and (2) 'that what we owe to others does not intrinsically (i.e., non-instrumentally) depend on such factors as nationality, citizenship, race, ethnicity, cultural and religious affiliation and the like' (Holtug 2011, 147–148).

The current debates have shifted away from strict positions that hold, for example, that 'the duty to provide aid neither gets weighed against any extra duty to help locals or compatriots nor increases in strength when locals or compatriots are in question' (Kleingeld and Brown 2011) to *moderate* positions. Recent examples of moderate positions are 'layered cosmopolitanism' (Held 2010, Chapters 2–3), 'approximate moral cosmopolitanism' (O'Neill 2000, 201) and 'transnationalism with a cosmopolitan inflection' and 'moderate cosmopolitanism' (Scheffler 2001, 129).[1] Some strict cosmopolitans, such as Martha Nussbaum and David Held, have changed their understanding of cosmopolitanism and argue now in favour of moderate versions of cosmopolitanism (Brown and Held 2010, 155; Held 2010, 100).

In my account of normative principles, the cosmopolitan theory can be understood in two ways. Either as an abstract moral principle from which abstract ought-judgements are derived. As such, the abstract ought-judgements are normative principles that say something about *what* ought to be done but nothing about *who* ought to do it; or as a cosmopolitan position that holds that it is possible without further justification to derive concrete ought-judgements about *who* ought to do from the abstract normative principles about moral universalism.

Recall that according to the account presented earlier, the difference between abstract and concrete ought-judgements is whether merely abstract or also concrete facts are considered relevant for the definition of normative principles. With this in mind, we can differentiate between two models for cosmopolitan theory: one in which the abstract principles cannot identify concrete ought-judgements without a set of further concrete facts and one in which the abstract principles identify the abstract ought-judgements:

i   *Concrete model*: Abstract principles + concrete facts = concrete ought-judgements
ii   *Abstract model*: Abstract principles = abstract ought-judgements

Appreciating this distinction, the shift from strict to moderate versions of cosmopolitan theory is a shift from the second model concerning abstract ought-judgements to the first model concerning concrete ought-judgements. The goal of contemporary accounts of abstract moral principles defended within global justice debates is thus not only to engage in a moral philosophical debate of moral principles but also to contribute to political solutions to the global challenges of our time.[2] Good examples of cosmopolitan analyses of the first model that combine abstract normative principles with concrete claims are analyses of the World Trade

Organisation (Higgott and Erman 2010), taxation (Murphy and Nagel 2005), and carbon markets (Caney 2010; Caney and Hepburn 2011).

Cosmopolitanism and liberal nationalism

Now, let us apply the two models for ought-assignment to the contemporary debate on global justice, a long-standing debate in the global justice literature which has implications for the discussions on global environmental justice. There have been profound disagreements in this debate among proponents of cosmopolitanism and liberal nationalism about whether nations have a moral significance that allows for special and national responsibilities in contrast to global ones (Scheffler 2001; Tan 2004).

Many contrast liberal nationalism with cosmopolitan arguments for global justice. Moral cosmopolitans maintain, Garrett W. Brown and David Held argue, 'that there are moral obligations owed to all human beings *based solely on our humanity alone*, without reference to race, gender, nationality, ethnicity, culture, religion, political affiliation, state citizenship, or other communal particularities' (Brown and Held 2010, 1, emphasis added). In contrast, liberal nationalists attach ethical significance to group membership, which allows one to give priority to individuals belonging to one's group, nationality, or culture. For example, if two individuals need to go to the hospital, the individual belonging to one's own group will be given priority (D. Miller 1995, 65–66).[3]

Applying the fact-sensitive account of concrete ought-assignments in contrast to abstract ought-assignments allows for a re-interpretation of the disagreements between liberal versions of cosmopolitanism and nationalism. I argue in particular that the disagreement concerns less abstract ought-judgements than it concerns concrete ought-judgements. I note further that the disagreement is not primarily normative, but concerns methodological disagreements about what facts are relevant with regard to the determination of concrete ethico-normative principles as well.

Now, consider the fact-sensitive interpretation of cosmopolitanism and liberal national principles. When we distinguish between the abstract and concrete models of normative ought-judgements, we are interested in whether the principles focus on abstract facts or include concrete facts as well. As such, one conclusion is that abstract cosmopolitans focus on *general* facts about what human beings share, such as basic needs and cognitive and moral capacity, whilst liberal nationalists focus on *specific* facts about what humans don't share, such as nationality and cultural affiliation. In other words, I distinguish between (1) *general and abstract facts* about natural and human properties from (2) *specific and concrete facts*. Specific facts can be further subdivided into (1) culture-specific facts, (2) context- or institutional-specific facts, and (3) policy-specific facts.

Similarly, David Miller distinguishes 'political philosophy for Earthlings' from a 'Starship Enterprise' type of political philosophy (D. Miller 2008, 31). Whereas the Starship Enterprise takes no facts about the earth into account, political philosophy for Earthlings is defined as being 'sensitive not only to general facts about the human condition but also to facts of a more specific kind, facts about particular societies, or types of societies' (D. Miller 2008, 31). However, whilst Miller argued that moral cosmopolitans embrace a fact-free theory, this is not an adequate description according to the account developed earlier. What moral cosmopolitans do is to allow for general, but not concrete, facts in the construction of normative principles.

Another way to understand the difference is to apply what Nils Holtug (2011) called *argumentation from above* as opposed to *argumentation from below*. *An argumentation from below* takes its point of departure from lower-level facts such as shared public culture. *Argumentation from above* presumes that lower-level facts do not intrinsically matter, but that they may matter derivatively because they constitute higher-level facts (Holtug 2011, 153). Holtug argued,

> [I]f the relevant lower-level facts obtain between two people, it is not a *further* fact about them that they are co-nationals. This fact just consists in these other facts. The fact that they are (also) co-nationals is a conceptual fact, rather than a further fact about the relations between them.
>
> (Holtug 2011, 153–154, emphasis in original)

This statement concerns the question of how to define personal identity and whether certain facts (about the person) add further information about the reality – i.e., *further* facts about the person. We are concerned here about whether a given fact may be classified as a *conceptual* fact adding 'no further information about reality' but 'further information about our use of the words "persons"' (Parfit 2007, 32). Now the question is whether there is an agreement about which facts are further facts and which are conceptual facts.

In his discussion of the ethical relevance of nationality, Holtug argued that national and social relationships do not add further information about personal identity. Nationality and national identity are lower-level facts that add no further information about the personhood, and nationality holds no *sui generis* level of identity. Similarly, Nussbaum argued that the list of different cultural and personal particularities is one of morally arbitrary attributes that are secondary to the person's moral worth as a human being (Nussbaum 2010, 29). I note that this description is not a contribution to the ethico-normative question of who ought to do what, but it does contribute to the definition of abstract moral principles. So conceived, the lower-level facts of nationality and culture have no implications for abstract principles.

The indeterminacy challenge

A group of cosmopolitans, including Martha Nussbaum and David Held, slide from the discussion of abstract moral principles to the institutionalisation of concrete ought-judgements. However, according to the distinction between abstract and concrete models of cosmopolitanism, their accounts of institutionalisation fail because of what I call an *indeterminacy challenge*. This challenge occurs when one defends a set of abstract moral principles that are defined independently of concrete and lower-level facts but nonetheless are used to embrace or critique particular institutionalisations of concrete ought-judgements.

The indeterminacy challenge contains two notions. On the one hand, it is commonly reckoned that 'moral cosmopolitanism makes no necessary *institutional* demands or recommendations' (Tan 2004, 94, emphasis in original). A dichotomy between the moral and political realm is embraced by defining moral principles in this way. On the other hand, cosmopolitan norms are used to guide, evaluate, and critique the political and institutional realm. Thus, Thomas Pogge and others argued very reasonably that moral cosmopolitanism has moral implications for institutional schemes and human behaviour. They argued that

> [m]oral cosmopolitanism simply says that the individual is the ultimate unit of moral worth and that *how we ought to act* or *what kinds of institutions we ought to establish* should be based on an impartial consideration of the claims or each person who would be affected by our choices.
>
> (Tan 2004, 94, emphasis in original)

The argument goes like this: because (1) human beings share the same humanity and (2) what we owe to others is independent of contingent factors (such as nationality), (3) 'we are required to respect one another's status as ultimate units of moral concern – a requirement that imposes limits upon our conduct and, in particular, upon our efforts to construct institutional schemes' (Pogge 1992, 50).

The question is now whether the moral principles can adequately determine how to construct institutions. This is the general term of the *indeterminacy challenge*. Let us now consider one concrete example that criticises a world state and defends a multi-layered institutionalisation. I am aware that the current debate has shied away from discussions on whether a world state is attractive. One reason for this is that few embrace a world state or federation (Caney 2005a, 266). Nonetheless, it is worth examining the grounds on which the rejection is justified. With regard to the critique of a world state, O'Neill argued, 'Given the dangers of concentrating too much power in a borderless world, there would be many risks in a total abolition of boundaries' (O'Neill 2006, 200). Similarly, Nussbaum argued,

> [A] world state [. . .] is far from desirable. Unlike domestic basic structures, a world state would be very unlikely to have a decent level of accountability to its citizens. It is too vast an undertaking, and differences of culture and language make communication too difficult. The world state is also dangerous: if it should become unjust there is no recourse to external aid. Moreover, even if those problems could be overcome, there is a deep moral problem with the idea of a world state, uniform in its institutions.
>
> (Nussbaum 1996, 213–214)

Moreover, Held endorsed a multi-layered cosmopolitanism with an institutional mix:

> Within the cosmopolitan framework [. . .] the political authority of states is but one aspect of a complex, overlapping regime of political authority; legitimate political power in this framework embeds states in a complex network of authority relations, where networks are regularized or patterned interactions between independent but interconnected political agents, nodes of activity, or sites of political power.
>
> (Held 2010, 100)

Whilst O'Neill, Nussbaum, and Held presented reasonable arguments, the question remains: on what normative grounds can they reject a world state and suggest a multi-layered institutional mix? If moral cosmopolitanism is a theory of abstract principles – e.g., of universalism and individualism, it cannot according to the account I have defined here decide which concrete institutional design is the most attractive. Thus, the arguments against the world state and for the multi-layered institutionalisation are underdetermined because the moral principles cannot determine who ought to do what at what institutional level. In order to conceptualise the latter set of questions, we need concrete normative principles that combine moral principles with concrete facts.

I conclude that cosmopolitan arguments which use abstract moral principles to determine concrete normative principles for who ought to do what at what institutional level fail the *indeterminacy challenge*. They do not explicate the relationship between moral norms and political principles sufficiently because they have no concrete normative theory for why one institutionalisation may be preferable to another one. The concrete normative theorisation of which facts about institutions and politics are relevant because the abstract moral principles are too general to justify a specific institutionalisation.

The non-ideal account

Moderate cosmopolitans have tried to solve the indeterminacy challenge by relaxing the abstract cosmopolitan norms to *moderate* and *non-ideal* ones. Moderate cosmopolitans relax the strict cosmopolitan claim 'that the duty to provide aid neither gets weighed against any extra duty to help locals or compatriots nor increases in strength when locals or compatriots are in question' (Kleingeld and Brown 2011). For example, the instrumental value of the nation-state is accepted. In this regard, moderate cosmopolitans and moderate liberal nationalists can be reconciled as compatible (Tan 2004, Chapter 5). The non-ideal approach does not relax abstract moral principles, but lowers the expectations of what people can or will provide. The non-ideal cosmopolitan embraces moral universalism and individualism, but relinquishes full implementation thereof. The non-ideal approach is considered consistent with the cosmopolitan ideal. '[A] world in which some fundamental principles or justice govern relations between all persons in all places' is maintained, even though the ideal is not expected to be fulfilled (McKinnon 2005, 248).

Nonetheless, the non-ideal model cannot be a solution to the indeterminacy challenge. Lowering the expectation of fulfilment of abstract norms (due to lack of institutional capacity, political willingness, or personal motivation) does not solve the indeterminacy challenge. It is still abstract principles that evaluate the behaviour and acts of political and institutional bodies. The non-ideal implementation model maintains that the goal is to implement those abstract moral principles in the political realm. As shown earlier, abstract moral principles face difficulties in offering concrete normative grounds for deciding between different kinds of non-ideal political institutions and practices. What kinds of non-ideal circumstances can be justified? Which kinds of non-ideal implementation of fact-free moral principles can be accepted and considered compatible with moral cosmopolitanism?

Abstract moral principles are too general to offer an account of which institutionalisation might be preferable. Thus, I conclude that the non-ideal approach fails to solve the challenge, even though it recognises the challenge of indeterminacy and the need to shift from abstract cosmopolitanism to concrete normative theorisation.

The contextualist model

Let us now consider another alternative – namely, what I call a *contextualist* model. I argue that the *contextualist* model for constructing concrete normative principles succeeds in solving the indeterminacy challenge, but risks failing on other grounds. In the strict and moderate models, politics, and institutional practices are obstacles to the realisation of moral principles; it takes a great effort to implement and institutionalise abstract moral principles in the political and social realm. In the contextualist model, the political and social realm is not perceived as an obstacle. On the contrary, specific facts about culture, institutions, or social contexts are built into the construction of concrete norms. Contextualist scholars, such as David Miller and Andrea Sangiovanni, criticise the strict and moderate cosmopolitan theories for a *dichotomist* model, which remains

> blind to the underlying and sustaining structures that make the pursuit of justice both possible and necessary. By trying to keep the realm of 'morality' unsullied by the demands of politics (which only become relevant after the philosopher has done his labour), the practice-independent theorist risks seeing institutions and the people participating in them as mere obstacles in the way of this ideal. He will tend either to overestimate the capacity of human beings to transform their condition through political action in controllable and predictable ways, or come to resent his fellow for not trying.
>
> (Sangiovanni 2008, 158)

According to Miller, abstract political philosophy resembles the idea of the Starship Enterprise:

> The Starship Enterprise view draws a line between political philosophy proper, which involves defining concepts and setting out principles in an entirely

fact-free way, and applied political theory, which takes these basic concepts and principles and, in the light of empirical evidence, proposes a more concrete set of rules to govern the arrangements of a particular society, or a particular group of societies.

(D. Miller 2008, 31)

Based on my fact-sensitivity account, contextualist scholars are right in critiquing the dichotomist model for failing to construct concrete normative principles with specific facts about institutional practices and politics. Let us consider whether the contextualist model provides a more convincing alternative. Two contextualist arguments can be distinguished. The first is that concrete normative principles and rules of regulation are context-dependent. One example of context-dependency is the so-called *practice-dependence thesis*, which says that '[t]he content, scope and justification of a conception of justice depends on the structure and form of the practices that the conception is intended to govern' (Sangiovanni 2008, 137–138).

According to Sangiovanni, the reason why justice is necessary as a societal principle of fairness is caused by common-sense knowledge of how the world and societies are organised. If political societies were completely different from the current ones, questions of fairness and justice might not be relevant topics of political philosophy. The need for justice and fairness is revealed by the study of social and political injustice and unfairness:

> [When] elaborating a conception of justice, we must first understand the role that the concept is meant to play in political action and criticism. Without some idea of this role, the journey from concept to conception would be traveled without a compass. To understand this role, in turn, we must seek an account of how it emerges within specific, institutionally mediated political and social contexts.
>
> (Sangiovanni 2008, 164)

In other words, the contextualist model appreciates that a positive conceptualisation of institutions is needed in order to comprehend the role of justice and fairness in political and societal realms. Practices and contexts condition the formulation and justification of concrete normative principles. Without practices and political institutions, there would be no realisation of concrete normative principles.

The second contextualist argument is more controversial: it says that moral and abstract principles are also context-dependent. The idea is that 'existing institutions and practices play a role in the basic *justification* and *formulation* of first principles' (Sangiovanni 2008, 137, emphasis in original). This means that the justification and formulation of moral principles of justice is not a practice-independent task but is instead practice- and context-dependent. Miller made a similar argument:

> Even the basic concepts and principles of political theory are fact-dependent: their validity depends on the truth of some general empirical propositions about human beings and human societies, such that if these propositions were

shown to be false, the concepts and principles in question would have to modified or abandoned.

(D. Miller 2008, 31)

Moral principles would not be formulated or even necessary if it were not for the relevant contexts challenged by injustice. According to Sangiovanni, this justifies the claim that institutional practices are *constitutive* of concrete normative principles. He argued that the practice-dependence justifications of institutions and practices are the reasons why we need principles of justice to guide the political and social realms in the first place.

> This account [. . .] is provided via an interpretation of both the point and purpose of the institutions for which the principles are needed, and the relations among participants in those institutions. It is only with these interpretative materials in hand, the institutionalist argues, that we can fill out the content of a particular conception of justice.

(Sangiovanni 2008, 164)

Whilst Miller focuses on contextualism related to the cultural and social realm, the type of contextualism that Sangiovanni defends takes an institutional form. Instead of focusing on societal cultures, he believes that moral contexts of justice vary according to different institutional contexts. Nation-states are one type of institutional setting. Another example is international institutional regimes, such as the European Union, the World Trade Organization, and so on (Sangiovanni 2008, 146). According to Miller, people are subjected to two sets of justice requirements concomitantly (D. Miller 2007, 44). This can be interpreted in two ways. One way to understand the argument is that Miller claims that contextualism applies to both (1) the political/social context and (2) the moral context, and that each societal sphere arguably reflects one type of moral context. One implication of this claim would be that people from different social and political contexts do not share the same moral context. This interpretation would lead to cultural and moral relativism.

Another interpretation could be to argue that the contextualist claims only apply in the political and social context, not the moral context. This interpretation calls for a closer look.

The split-level model

As we have seen, the dichotomist avoids a naturalistic fallacy, but risks failing on an indeterminacy challenge. The contextualist model solves the indeterminacy challenge by contextualising normative concrete principles for societal justice and abstract principles of justice. However, this model risks failing on moral relativism by allowing for different moral contexts. This would be incompatible with the moral universalism and individualism of cosmopolitan theory.

In contrast, when we apply the fact-sensitive account of ought-assignments, we distinguish sharply between theorisations that concern moral principles on the one hand and theorisations that concern concrete ought-assignments and institutionalisations of normative principles on the other. As a result, we allow for a consistent interpretation of the contextualist model that avoids the risk of moral relativism. The first condition for this is that the dichotomy between the moral and political realm is appreciated. This follows the logic of distinguishing between principles of justice and principles of personal morality (Rawls 1958, 164), and distinguishing between rules of regulation and fundamental moral principles (Cohen 2008). The notion is that principles of justice are concrete normative principles that should not be conflated with abstract principles of morality. Justice and fairness apply in societal institutions and background conditions, whilst principles of morality concern interpersonal moral obligations (Pogge 1992; Broome 2012). The institutional theory of justice is composed as a principle 'to specify the fair terms of social cooperation' (Rawls 2003, 7).

By distinguishing between morals and justice, I suggest a *split-level model* as an alternative to the dichotomist and contextualist models, but which nonetheless combines insights from the dichotomist and contextualist models. The split-level model takes a contextualist approach to the concrete principles of justice and ought-assignments, but avoids moral relativism by maintaining the dichotomy between concrete principles of justice and abstract morality.

Similarly, Pogge distinguishes institutional requirements from interactional ones depending on whether we apply fundamental cosmopolitan principles of justice to institutional schemes or fundamental cosmopolitan principles of ethics to human interactions (Pogge 1992).[4] Taking Pogge's distinction into account, we can distinguish two cosmopolitan areas of concern, one in which abstract moral norms are applied to ethical principles for human behaviour and interaction and one in which abstract moral norms are applied to political principles for global distributive justice. Whilst the former argument remains within the area of moral philosophy (because only facts about humanity are presumed), the latter aims to reconcile insights from moral philosophy with insights from normative political theory and the social sciences.

In the conceptualisation of justice, which focuses on institutions, individuals are not morally responsible for doing injustice to other particular persons.[5] Individuals are responsible for doing good, for example, by avoiding harming the climate and other people (questions of goodness), but this is not a question of justice. In the institutionalised conceptualisation of justice, individuals are important agents of co-responsibility, but the responsibility for maintaining a just social cooperation cannot be reduced to responsibility held by the single individual. This is because individuals' just or unjust relationship is institutionally mediated. Justice depends on a concrete normative theory of how institutions foster just relationships.

The split-level model and liberal nationalism

Several passages in the works of Miller and Sangiovanni indicate that they follow the split-level model. For example, Miller distinguishes between a universalist and a contextualist component. The *universalist* component manifests the principles of global norms that are applicable in every context. This applies what Miller calls 'equal moral concern' for every human being (D. Miller 2007, 43–44). In contrast, the contextualist component identifies different contexts, such as family, universities, and nations, where different principles of justice apply.

> Rather than laying out principles P1 . . . Pn as constituting justice in all circumstances, it should take the form 'In C1, P1; in C2, P2; . . . Cn, Pn', where the Cs are the distributive context in which principles of justice may be applied.
>
> (D. Miller 2001, 7)

This means that each context contains a particular set of concrete principles of justice. There exists no single principle of justice that can be implemented everywhere, independent of context. According to Miller, what is special about this type of contextualism is that 'the contextually specific principles are free-standing, and not simply derivatives of invariant basic principles' (D. Miller 2001, 7). This way of understanding justice differs from the universalist position that applies the same set of universal principles of justice in every context. Instead of aiming to implement or realise one set of norms, the contextualist aims to apply a different set of principles representing 'the best local application of the fundamental principles, which hold everywhere' (D. Miller 2001, 24).

The split-level approach is a dichotomist model because it maintains that abstract universal moral principles exist, but it claims that these principles cannot determine concrete and practical implications and distribution of obligations and relevant associated tasks. The implication of this split-level model between the moral and the societal context is that the demands one may assign individual agents depend on what context is in question.

In the fact-sensitive account, the split-level model can be appreciated because moral principles of global justice and principles of what ought to be done underdetermine the concrete normative principles of institutionalisation as well as the concrete ought-assignments. Thus, what remains is to consider the conceptualisation of the concrete ought-assignments.

Let us continue with the example of liberal nationalism. In the split-level model, proponents of cosmopolitanism and liberal nationalism share several ideas about fundamental norms despite the fact that they disagree about what institutionalisation should be prioritised.[6] Indeed, liberal nationalism and the ideas of the ethical relevance of nationality are parasitic to fundamental abstract fundamental liberal ideas about individualism and moral universalism. If liberal nationalism did not embrace moral universalism, it would slide towards moral or cultural relativism, which it does not. Liberal nationalists claim neither that there is no common human nature nor that every person does not enjoy equal moral status (Caney 2005a, 39).[7]

Note that 'the broad empirical assumption that there exists, among many peoples, a sufficient degree of convergence in attitudes and beliefs', which identifies a national identity (D. Miller 2007, 124–126) is fully compatible with embracing moral universalism. One can coherently embrace concrete fact-sensitive norms that national communities are instrumentally or even intrinsically valuable for their members and their respective identities and at the same time that people around the world share the same morally relevant properties.

Similarly, it is conceptually possible for someone to endorse cosmopolitanism and defend the instrumental or intrinsic value of nation-states for their members. The latter is what Bader refers to as 'value-independent, intrinsic justification of associative relations and belongings [. . .] by the respective actors themselves' (Bader 2005, 87). In fact, it would be a categorical mistake to move from the ideas of cultural identity to a rejection of the moral principles applying to all persons with common morally relevant properties. The validity and plausibility of liberalism defined by individualism and moral universalism are rooted in arguments independent of these anthropological assumptions.

Despite the agreement about people's equal moral status, there is considerable disagreement among liberal cosmopolitans and nationalists, which justifies the distinction between the two. The disagreement can be determined in three parts: first, they may disagree about the normative and political relevance of nation-states. Here, liberal nationalists stress the ethical significance of nation-states more strongly than liberal cosmopolitans. Second, they may balance societal and political concerns differently, primarily when they weigh global justice versus social justice at the global and national levels. Hence, even though they agree on the significance of both global and social justice, they do not necessarily agree on how to distribute resources between the two. Third, liberal cosmopolitans and liberal nationalists determine and justify policy by pointing to different topics. Liberal nationalists draw attention more vigorously to the importance of cultural communities and national commitments in determining and justifying different policy decisions. Conversely, liberal cosmopolitans tend to take a juridical approach when they endorse top-down policy decisions, such as transnational rights.

Levels of institutionalisation

Having argued that concrete normative principles for institutionalisation of moral principles and concrete ought-assignments require a theory of what political or social context should be considered relevant, let me now turn to the specific discussion of what level of institutionalisation might be embraced and for what reason. Which sets of facts that are politically relevant remains contested: it continues to be an area in need of political theoretical scrutiny. There is no agreement on whether concrete normative principles should be sensitive to specific facts about nationality and national identity.

As we have seen, some claim that nationality constitutes a morally significant characteristic of a person's identity, culture, or community (D. Miller 2007, 30–33). According to the split-level model, this claim can be conceived as a justification of why national political bodies as collective agents are relevant in cosmopolitan theory as factual circumstances for global politics. Miller defends this as an example of what he calls *politically relevant* normative principles:

> It is tempting to assume that those who would live for 400 years would be just like us, but with a much larger capacity for welfare; on reflection, this must obviously be wrong. Their lives are simply different from ours. If such beings were to come into existence, this would radically disrupt our understanding

of the human form of life, and might lead us to conceive of human needs differently.

(D. Miller 2011, 171)

In the fact-sensitive account of concrete normative principles that I propose, it is indeed essential to take the point of departure from the current circumstances of politics and arguably not from hypothetical ideas of how the world might look like if people lived for 400 years.

Nonetheless, it remains to be discussed on what principle we may decide the level of institutionalisation of ought-assignment. Proponents of liberal nationalism have suggested that one politically relevant fact is the distinction between global and national spheres of politics, and particularly that the national level sustains its relevance because of the cultural identities attached to the national level. Since nation-states still play a significant role in modern national and global politics, it might be justified to consider the facts about nation-states that are necessary for the construction of concrete normative principles for ought-assignments.

I note, however, that the argument that the institutions of nation-states maintain political relevance is logically independent of the arguments for national identities. Thus, we can differentiate between two arguments: that nation-states are relevant political contexts when determining concrete ought-assignments because of the national identities attached to them and that nation-states are relevant political contexts when determining concrete ought-assignments because of what I later call their *institutional capacity*.

Note that here liberal nationalism is not conceived as an abstract normative principle for global justice, but as a concrete principle for what level of institutionalisation should be embraced. By elevating the facts about cultural and national identities, Miller proposes that nationality is a *further* fact about humanity, not just a *conceptual* fact. Recall that a further fact adds information about the reality, whereas a conceptual fact only adds information about how we use the words.

One way to elaborate on this distinction could be to perceive the question of *further* facts as a methodological issue.[8] If one presumes – e.g., a culturalist methodology – one can claim that cultural and national identities are further facts about humanity. In contrast, if one embraces an individualist or institutionalist methodology, one would say that cultural and national identities are merely conceptual facts. What matters here are the institutions. By so arguing, specific conceptualisations of liberal nationalism should not be rejected on normative reasons. Yet the conceptualisations can be criticised for presuming an implausible methodology and arguably a wrong set of facts. Note that the arguments for liberal nationalism might nonetheless be *theoretically consistent*, which means that the set of concrete normative notions cannot be criticised if they consistently follow from the set of methodological and meta-theoretical assumptions. In contrast, the critique should be targeted at the level of the fundamental principles and methodological assumptions that lead to the normative claims.

In conclusion: we cannot answer the fact-sensitive question of what level of institutionalisation of ought-assignment should be embraced without first deciding what meta-theoretical and methodological assumptions can be justified. In the next chapter, I continue this discussion and defend an institutionalist methodology that rejects the collectivist approaches, such as culturalism.

Conclusion

The purpose of this chapter has been to discuss the indeterminacy challenge of abstract normative principles. I have argued that abstract principles fail to determine what political institutional level might be preferable, and thus fail to determine who ought to do what. I have argued in favour of a split-level approach in which abstract normative principles are maintained, whereas which social context is politically relevant is partly answered by concrete facts.

The debate about cosmopolitanism and liberal nationalism can be reconstructed in light of the distinction between abstract and concrete normative theories. One cause of their different views is a meta-theoretical disagreement about whether general facts or specific facts are relevant in the construction of concrete normative principles. So conceived, the main line of conflict is not the disparity between cosmopolitanism and nationalism, but between abstract and concrete approaches to political philosophy that transform the global justice debate from a normative discussion about the moral arbitrariness of nation-states to a meta-theoretical and methodological debate about what facts are considered relevant for normative theoretical arguments for concrete ought-assignments.

Notes

1 Some of the distinctions are mutually compatible; others are not. For further clarification of cosmopolitanism, see, e.g., Scheffler (2001) and Tan (2004).
2 Contributions on cosmopolitanism and global justice are generally concerned with global poverty. See, e.g., Singer (1972), Carens (1987), and Pogge (2005). For global climate change, see, e.g., Caney (2010) and Shue (2010). For border, territorial rights, and immigration issues, see, e.g., Scheffler (2001), Miller (2007), and Pevnick (2011).
3 Liberal nationalism defends national self-determinacy. For non-culturalist arguments for national self-determinacy, see, e.g., Buchanan (2004), Caney (2005a, Chapter 5), and Pevnick (2011).
4 Interactional cosmopolitanism focuses on individual moral obligations. It can be subdivided into utilitarians, such as Singer (1972) and Unger (1996), and a broad group of Kantians, such as Nussbaum (1996), O'Neill (2000), and Benhabib (2004). Whilst the former group draws attention to individual responsibility for harming others, the latter group of scholars emphasises personal duties to enhance civic patriotism and provide universal hospitality and global citizenship. Conversely, institutional cosmopolitanism focuses on the shared responsibility people may have to contribute to political institutions that fulfil normative objectives (Stemplowska 2009).
5 Few have addressed the normative-institutional aspect of global justice. Important exceptions are Rawls (1993, 1999, 2001), Miller (2007), and Vanderheiden (2008).
6 Similarly, Pogge argued that Miller's concept of contextualism (defined by the claim that different fundamental moral principles apply in different contexts) overlapped with moral universalism (Pogge 2002, 37–39).
7 See Singer for examples of thinkers who reject the idea of a single moral standard (Singer 2002, Chapter 5). Singer himself has argued that the moral status of sentient

creatures cannot be limited to human beings, but should also include sentient animals (Singer 1993).
8 For discussions of collectivism, supervenience, and methodology, see Chapter 3. Note that in this chapter, we have primarily been interested in the meta-theory of culturalism and institutional theory. Meta-theory and methodology are closely intertwined. Meta-theory focuses on what objects, rules, and principles are acceptable in a scientific discipline. Methodology focuses on what methods are acceptable to obtain knowledge about objects, rules, and principles in a scientific discipline.

References

Bader, V., 2005. Reasonable impartiality and priority for compatriots. A criticism of liberal nationalism's main flaws. *Ethical Theory and Moral Practice*, 8 (1–2), 83–103.

Benhabib, S., 2004. *The rights of others: aliens, residents and citizens*. Cambridge: Cambridge University Press.

Broome, J., 2012. *Climate matters: ethics in a warming world*. London: W.W. Norton.

Brown, G. W. and Held, D., 2010. Editors introduction. *In:* G. W. Brown and D. Held, eds. *The cosmopolitanism reader*. Cambridge: Polity, 1–14.

Buchanan, A., 2004. *Justice, legitimacy, and self-determination: moral foundations for international law*. Oxford: Oxford University Press.

Caney, S., 2005a. *Justice beyond borders: a global political theory*. Oxford: Oxford University Press.

Caney, S., 2005b. Cosmopolitan justice, responsibility, and global climate change. *Leiden Journal of International Law*, 18 (4), 747–775.

Caney, S. and Hepburn C., 2011. Carbon trading: unethical, unjust and ineffective? *Royal Institute of Philosophy Supplements*, 69 (October 2011), 201–234.

Carens, J. H., 1987. Aliens and citizens: the case for open borders. *The Review of Politics*, 49 (2), 251–273.

Cohen, G. A., 2008. *Rescuing justice and equality*. Oxford: Oxford University Press.

Held, D., 2010. *Cosmopolitanism: ideals and realities*. Cambridge: Polity.

Higgott, R. and Erman, E., 2010. Deliberative global governance and the question of legitimacy: what can we learn from the WTO? *Review of International Studies*, 36 (2), 449–470.

Holtug, N., 2011. The cosmopolitan strikes back: a critical discussion of Miller on nationality and global equality. *Ethics and Global Politics*, 4 (3), 147–148.

Kleingeld, P. and Brown, E., 2011. Cosmopolitanism. *In:* E. N. Zalta, ed. *The Stanford Encyclopaedia of Philosophy*. Spring 2011 edn. Available from: http://plato.stanford.edu/archives/spr2011/entries/cosmopolitanism/ [Accessed 3 May 2016].

McKinnon, C., 2005. Cosmopolitan hope. *In:* G. Brock and H. Brighouse, eds. *The political philosophy of cosmopolitanism*. Cambridge: Cambridge University Press, 234–248.

Miller, D., 1995. *On nationality*. Oxford: Oxford University Press.

Miller, D., 2001. Two ways to think about justice. *Politics, Philosophy and Economics*, 1 (1), 5–28.

Miller, D., 2007. *National responsibility and global justice*. Oxford: Oxford University Press.

Miller, D., 2008. Political philosophy for earthlings. *In:* D. Leopold and M. Stears, eds. *Political theory: methods and approaches*. Oxford: Oxford University Press, 29–48.

Miller, D., 2011. On nationality and global equality: a reply to Holtug. *Ethics and Global Politics*, 4 (3), 165–171.

Murphy, L. and Nagel, T., 2005. New Ed edition (18 November 2004). *The myth of ownership: taxes and justice.* Oxford: Oxford University Press.

Nussbaum, M., 1996. Patriotism and cosmopolitanism and reply. *In:* J. Cohen, ed. *For love of country.* Boston, MA: Beacon Press.

Nussbaum, M., 2010. Kant and cosmopolitanism. *In:* G. W. Brown and D. Held, eds. *The cosmopolitanism reader.* Cambridge: Polity, 27–44.

O'Neill, J., 2006. Knowledge, planning, and markets: a missing chapter in the socialist calculation debates. *Economics and Philosophy*, 22 (1), 55–78.

O'Neill, O., 2000. *Bounds of justice.* Cambridge: Cambridge University Press.

Parfit, D., 2007. *Is it personal identity that matters?* The Ammonius Foundation, December 31.

Pevnick, R., 2011. *Immigration and the constraints of justice: between open borders and absolute sovereignty.* Cambridge: Cambridge University Press.

Pogge, T., 1992. Cosmopolitanism and sovereignty. *Ethics*, 103 (1), 48–75.

Pogge, T., 2002. Moral universalism and global economic justice. *Politics, Philosophy and Economics*, 1 (1), 29–58.

Pogge, T., 2005. A cosmopolitan perspective on the global economic order. *In:* G. Brock and H. Brighouse, eds. *The political philosophy of cosmopolitanism.* Cambridge: Cambridge University Press, 92–109.

Rawls, J., 1958. Justice as fairness. *The Philosophical Review*, 67 (2), 164–194.

Rawls, J., 1993. *Political liberalism.* Cambridge, MA: Harvard University Press.

Rawls, J., 1999. *The law of peoples: with the idea of public reason revisited.* Cambridge, MA: Harvard University Press.

Rawls, J., 2001. *Justice as fairness: a restatement.* Cambridge, MA: Harvard University Press.

Rawls, J., 2003. *Justice as fairness: a restatement.* Cambridge, MA: The Belknap Press of Harvard University Press.

Sangiovanni, A., 2008. Justice and the priority of politics to morality. *The Journal of Political Philosophy*, 16 (2), 137–164.

Scheffler, S., 2001. *Boundaries and allegiances: problems of justice and responsibility in liberal thought.* Oxford: Oxford University Press.

Shue, H., 2010. Global environment and international inequality. *In:* S. M. Gardiner, S. Caney, D. Jamieson, and H. Shue, eds. *Climate ethics: essential readings.* Oxford: Oxford University Press, 101–111.

Singer, P., 1972. Famine, affluence, and morality. *Philosophy and Public Affairs*, 1 (3), 229–243.

Singer, P., 1993. *Practical ethics.* Cambridge: Cambridge University Press.

Singer, P., 2002. *One world.* New Haven, CT: Yale University Press.

Stemplowska, Z., 2009. On the real world duties imposed on us by human rights. *Journal of Social Philosophy*, 40 (4), 466–487.

Tan, K., 2004. *Justice without borders: cosmopolitanism, nationalism, and patriotism.* Cambridge: Cambridge University Press.

Unger, P., 1996. *Living high and letting die: our illusion of innocence.* Oxford: Oxford University Press.

Vanderheiden, S., 2008. *Atmospheric justice: a political theory of climate change.* Oxford: Oxford University Press.

3　Fact-sensitive ought-assignments

We have discussed that it is convenient to conceive normative principles in two ways. Abstract principles are fundamental normative principles that depend only on highly abstract facts, such as the fact that there exist human beings or that climate change is actually happening. On the other hand, concrete principles are principles that depend on a broader set of concrete facts about politics, states, and social engagement, such as people's nationality or cultural heritage. Until now, I have primarily defined abstract and concrete normative principles and offered arguments. First, concrete normative principles avoid sliding towards moral relativism and political realism because they are justified by fundamental normative principles. Second, we need these concrete principles because abstract principles fail to determine who ought to do what.

Although abstract principles provide a framework for what ought to be done, they do not fully determine which agents ought to what and at what institutional level. The task now is to distinguish abstract ought-judgements from concrete ought-judgements. As we know, we have sound moral reasons to believe (1) that *climate change is bad*. This axiological principle should be distinguished from (2) the fundamental normative principle that *something ought to be done* about it and, moreover, from (3) the concrete normative principles about *who ought to do* it.

We are concerned in this chapter with the normative criteria for the *can* and *fitness*-conditions under which particular agents can be held morally responsible for climate change. Can and fitness-conditions of agency relate to the discussion of whether *ought implies can*, which goes beyond the dichotomy between *is* and *ought*. As we have seen, a convincing approach to the is-ought discussion is that it is possible to maintain a logic dichotomy between norms and facts whilst arguing the dichotomy gets messier when we evaluate or describe social scientific objects. We may distinguish norms and facts as analytical categories, but for pragmatic, epistemic, and scientific purposes, statements about norms and facts cannot be completely disentangled. I propose that judgements about who ought to do what belong to this group of statements because the normative and factual components of the normative judgements about who ought to what are contingent on several theoretical and methodological premises which draw on both normative and factual content. These normative judgements contain both fundamental normative principles and factual components. Together, they provide an account of fact-sensitive ought-assignments.

This discussion suggests a link between the meta-theoretical discussions in Part 1 and the subsequent normative discussion on moral responsibility for climate change in Part 3. Thus, the purpose of this chapter is twofold. First, I offer an account of concrete ought-entailments which I later shall use in the discussion of moral responsibility for climate change. I argue that concrete ought-entailments enclose a set of normative criteria for which factual *can-conditions* can be used in the fulfilment of ought-statements. Second, I show that the normative criteria for can-conditions depend on abstract norms. We need to first discuss how to conceptualise ought-entailments, which leads to my argument that the concrete fact-sensitive account of ought-judgements is contingent on can-conditions/fitness-conditions, which nonetheless provide an insufficient account of normative ought-assignments that require normative criteria as well.

Ought-judgements

This section is concerned with the question of how to propose judgements about *who* ought to rectify the wrong-doings of climate change. As we know, there are several moral reasons for why it is wrong to contribute to environmental harm and climate change. Normally, we take a wrongdoing for saying something about what ought to be done. Consider one example: *if agent X is in pain, then X ought to be assisted* (Cohen 2008, 249). Similarly, assuming climate change is bad, we can say that *if X is harmed by climate change, X ought to be assisted.*

A Humean critique would state that this way of arguing fails on a naturalistic fallacy because an *ought*-statement is derived from an *is*-statement. To see why this is not the case, we need to invoke the earlier note that the use of certain descriptive judgements contains a normative evaluation of what is being described. In the *pain*-example, nothing is said about *what it is* that establishes a criterion for what ought to be done. The sentence makes a more modest point: when we say that one is in pain, we imply at the same time that the person in pain should be assisted. Note that the normative reason to assist the agent in pain is not contained in the fact that the agent is in pain, but in valid fundamental principles, such as *one ought to help people who are in pain* (Cohen 2008, 250).

As argued earlier, these fundamental principles about what ought to be done do not determine *who* ought to do what ought to be done (Sinnott-Armstrong 1984, 250; R. Stern 2004, 47). There might be simple cases, such as A ought not to kill B, in which what ought to be done overlaps clearly with what A ought to do (or rather, not do). However, here we are interested in the cases that are complex cases of global challenges. Complex cases are defined by moral principles for what ought to be done do not determine who ought to do what. I take climate change to be a complex case of normativity, which has implications for the theory of moral responsibility for climate change. How do we determine *who* is morally responsible in contrast to *what* is morally wrong?

By invoking the *ought implies can* (OIC), we say that ought-statements about who ought to do what are, among other criteria, dependent on what the agent *can do*. The question remains: what criteria for these can-constraints can be morally accepted?

Ought implies can

The principle 'OIC' links the capacities that the agent actually has with the normative requirements that one may assign to the agent. Let us begin with a preliminary formation: an agent at a given time has the obligation to do only what the agent at the time has the ability and opportunity to do (Vranas 2007, 167). Conceptual necessity means that the normative ought-statement contains claims about the ability and opportunity of the agent to perform the act:

- X ought to be done;
- A is obliged to do what A can do; and
- (OIC) If A should do X A can do X.

Let us distinguish two different understandings of the argument. The first understanding is that ought-judgement *entails* judgements about factual constraints; the second is that ought-judgement *presupposes* judgements about factual constraints. The main difference between the two understandings is not terminological, but addresses two different, albeit related, topics. For reasons that will become apparent, the presuppositional argument is defined by presuming OIC to concern how to define fundamental principles in a fact-insensitive or fact-sensitive way, whereas the entailment argument takes OIC to concern the conditions of human agents and thereby the definition of who ought to be what. I conclude in favour of the latter version because it provides the most robust account of what role the can-conditions should play in normative political theory. Let us begin by considering the presupposition principle.

When OIC is interpreted along the lines of the presupposition principle, a weak and strong reading of this principle can be distinguished. The strong reading says that the can-conditions are 'a fundamental determinant of what is right and wrong' (R. Stern 2004, 45). This means that no act can be right if it is beyond human capacities to act in this way or wrong if it is beyond human capacities to avoid acting in this way (R. Stern 2004, 45). So conceived, what is right and wrong is constrained by human capacities; otherwise, it will not apply to human agency. David Miller provides an example of this: 'if the facts were not to hold – if human beings were no more capable of self-conscious choice than other animals – then the principle would not apply' (D. Miller 2008, 34). Two options are now available.

One version of the presuppositional argument rejects the distinction between abstract and concrete normative principles since it is assumed that all kinds of norms are dependent on factual considerations of the concrete can-conditions of the agents. The presuppositional argument does not concern the concrete ought-judgements, but provides claims that have implications for the determination of abstract principles.[1]

Alternatively, in order to provide some plausibility to the presuppositional argument, a possible way to interpret the argument would be that it is a thesis concerning merely fundamental moral principles, not concrete normative principles, and, arguably, not ought-assignments. Thus, the presuppositional thesis maintains some plausibility if it is assumed that it is only a thesis about how abstract fundamental moral principles presuppose abstract facts about humanity and nature. So, unlike the strong case, where the can-conditions determine what is right and wrong, the weak case only says, in the words of Miller, that 'if F does not obtain, we have no reason to assert P' (D. Miller 2008, 36). Miller explains this point:

> Imagine, then, somebody who is disposed to accept the liberty principle, but who cannot see the relevance to it of the fact of human self-consciousness. Such a person must presumably think that, other things being equal, the liberty principle should also apply to certain animals. Explaining to this person the relevance of F would involve pointing out how a cow, say, would not experience choice as valuable – would not be able to imagine being in a field other than the one she was currently in, and therefore would not be deprived, in a morally relevant way, by being denied a choice of fields. This is just to remind our interlocutor of some very familiar facts of human experience, and to show how it is those facts that bring principles like the liberty principle into play – if the fact were otherwise there would simply be no reason to propose such a principle.
>
> (D. Miller 2008, 36)

Although Miller embraces the presuppositional thesis, he is not defending the strong view that the fact that people have a consciousness determines what is right and wrong. What he says is that it would be pointless and serve no reason to propose a fundamental normative principle of humanity if it did not invoke knowledge about fundamental facts about humanity (R. Stern 2004, 49). In support of this view, Miller further argued,

> The principles political philosophers propose must be principles that citizens can act upon, not in the sense that they can fully implement them here and now, but in the sense that their present actions can be guided by the longer-term goal of realizing the principles in question.
>
> (Miller 2008, 44)

Nothing is said in this passage about what determines what is right and wrong. What is discussed is with what kinds of principles political philosophy we should be occupied.

Two preliminary conclusions should be stressed. The strong view of the presuppositional argument determines what is right and wrong but fails to invoke an account of what normative criteria of the can-conditions are justifiable. In contrast, the weak view defends what I earlier called abstract fundamental normative theory, which depends on a set of highly abstract facts, such as human consciousness. However, little is said about the can-conditions related to ought-assignments to particular agents. Moreover, little is said about how 'to judge the value of the act as right on its own merits, regardless of the capacities of particular agents' (R. Stern 2004, 51). Thus, I conclude that both the strong and weak views are unsatisfactory in providing a robust account of how to comprehend what role the can-conditions play in ought-judgements. Let us now turn to the *entailment* interpretation of the OIC principle.

Ought-entailments

In contrast to the presuppositional reading, which concerns how to define fundamental principles, the entailment reading focuses on the can-conditions that determine what agents – all things considered – ought to do at a specific time. Moreover, it sets limits on how much we can demand from particular agents. This means that the OIC implies the following principles (Howard-Snyder 2006, 234; Vranas 2007, 169–171):

> *Necessarily, if A ought to do X at T (i.e., has a pro tanto obligation), then A can do X at T.*

In order to make this argument work, we need to invoke the distinction established earlier between abstract fundamental principles on the one hand and concrete normative principles on the other. The fundamental moral principles about what is right and wrong are abstract principles that are only contingent upon abstract facts, whilst ought-entailments are concrete normative principles that are also contingent upon concrete facts. By so arguing, it is possible to determine what an agent ought to do contingent on what the agent can do – without claiming anything about the fundamental principles of what ought to be done. Put differently, the fundamental moral principles do not lose their normative content, although agents do not have the capacity to follow them (R. Stern 2004, 56). Robert Stern argued,

> The moral law only has its status of being obligatory for us because we are able to act upon it, and that we can thus only explain this obligatoriness by accepting certain claims about our capacities and their conditions ('we ought implies we can'). But this is distinct from the claim that no act can be right (rather than just obligatory for us) unless we are able to perform it.
>
> (R. Stern 2004, 57)

This means that the argument for ought-entailments does not imply that there could not be fundamental moral principles that are still valid even though human agents cannot imitate them (R. Stern 2004, 59). If it is appreciated that OIC only applies to concrete ought-statements, one important question remains. If we argue that what A ought to do is dependent on what A can do, is it then a normatively significant matter what A can do? As we have seen, many scholars presume a sharp division of labour between studies of factual and normative matters. In contrast, in the case of concrete ought-statements, the judgements about what A *can* do are deeply embedded in the judgements about what A *ought* to do. If this is correct, an important aspect of concrete normativity is the criterion for what A can do.

Inquiries of what A can do touch upon two types of questions. One type of inquiry of what A can do is a classic philosophical question of determinism versus moral responsibility, which is related to the question of compatibilism and incompatibilism. Compatibilism says that the matter of moral responsibility is not sensitive to the questions of determinism (Frankfurt 1969; Fischer and Ravizza 1998); incompatibilism argues that if determinism is true, then agents cannot be morally responsible for anything (Lippert-Rasmussen 2005b). Today, the debate also has a biomedical variant in which moral responsibility and agent's free will may be manipulated by medicine (Persson and Savulescu 2012).

Another type of question concerns the inquiry of what A can do as a social scientific question of explanation, causality, and agency. The answer to these questions is dependent on what methodological point of departure is assumed. The reason for this is assumed because social inquiries of explanation, causality, and agency are methodologically embedded theories (S. Miller 2010; Björnsson 2011; Vincent and van de Poel 2011). The philosophical and methodological inquiries raise the same set of questions but approach them differently. In the next section, I concentrate on the social scientific aspects of can-conditions because I am particularly interested in investigating the conditions which constrain agency in social and institutional contexts.

What can 'A' do?

What types of incapacity to act are normatively acceptable (Mason 2003, 319)? This question, which will be treated in more detail later, has implications for the discussion of moral responsibility for climate change. Depending on what necessary and sufficient criteria for can-conditions one accepts, the assignment of moral responsibility differs with regard to what types of agents are fit to be held responsible and for what reasons. Before proceeding, let us examine which criteria may be embraced. Note that this way of framing the debate on moral responsibility and ought-entailments does not fail on a naturalistic fallacy. No normative conclusion about what A ought to do is derived from factual matters about what A is capable of doing alone. What is questioned is which can-conditions of A's acts should be entailed in concrete ought-judgements of what people ought to do.

Let us begin by considering some of the reasons why OIC has been criticised for being implausible. One argument is that it is frequently people's own fault that they cannot do what they ought to do (Sinnott-Armstrong 1984). Here, the relevant disagreement concerns whether we allow that people's own fault is an acceptable criterion for can-conditions. To clarify the debate on the normative criteria for can-conditions, I distinguish between four different criteria based on two distinctions between (1) internal and external constraints and (2) soft and hard constraints. Hard constraints rule out can-conditions because of physical or biological impossibility. Soft constraints, on the other hand, differ from hard constraints in the sense that they do not rule out can but indicate how can-conditions can be comparatively less or more feasible (Gilabert and Lawford-Smith 2012, 813–815). For example, a hard constraint on agent A's possibility of jumping to the moon may be the agent's physical capacity, whilst the religious mind of agent A might be a soft constraint in the sense that it makes A less likely to get a divorce although it is not (physically or neurologically) impossible for A to do so.

The distinction between internal versus external criteria refers to whether the agent has an *external* pro tanto obligation to do X in the sense that the obligation corresponds to the agent's external situation, or whether the agent has an *internal* obligation in the sense that the obligation corresponds to the agent's justified belief about what to do (Vranas 2007, 169). The internal view on obligations questions whether people feel obliged or whether they intend to do something good or bad, whilst the external view draws attention to the external obligations one might have in a given context or factual circumstances. The outcome of these two distinctions is four categories. I have tentatively named the categories to illustrate what might exemplify the different categories. They should by no means be regarded as exhaustive (cf. Table 3.1).

It should be added that none of the four categories are unchangeable. Nano- and bio-technological developments, for example, may change the hard constraints on can-conditions (Gilabert and Lawford-Smith 2012, 813). Moreover, institutional developments may change the soft constraints. In contrast to the realists who, as we have seen earlier, take a static approach to the factual circumstances, proponents of institutionalism frequently draw attention to the *institutional resources* in contemporary societies, which have the capacity to change social facts.[2] Here, institutions can transform factual conditions and thereby act as a facilitator of norms (Sangiovanni 2008).

Having distinguished four types of can-conditions, let us now consider which ones we may allow to be normatively significant with regard to ought-judgements. For descriptive purposes, all four categories might be valuable for explaining why A did or did not do X. However, when we are interested in judgements about what A ought to have done, we ask a different set of questions. We are primarily interested in whether A can be held responsible for the outcome X or whether A can be

Table 3.1 Four types of constraints

	Internal	*External*
Soft	Psychological constraints	Social and political constraints
Hard	Physiological constraints (including neural capabilities)	Physical constraints

blamed for failing to do X. Thus, we are interested in the necessary and sufficient criteria for the can-conditions of moral responsibility.

We might start with the simple matter that one necessary criterion is that A is not physically constrained (Mason 2003, 321). Moreover, normal mental health and the absence of various kinds of neurological disorders are presumed to be a necessary criterion for agents to be capable of being held responsible. This means that both the internal and external variant of the hard constraints are necessary conditions for agents to be fit to be held responsible. A different way of making the same point is to say that these are necessary criteria for upholding a morally relevant agency.

On the other hand, the absence of hard constraints is not a sufficient reason to assign moral responsibility to A. Moral responsibility is not merely a question of what is physically, biologically, and logically possible (Mason 2003, 321). Normally, we do not without further justification blame A for not helping a person out of car crash that occurred in a different city, although no hard constraints would make it impossible for A to drive, let us say, two hours to help the person out of the car crash. If we would blame A in this case, there has to be a further normative reason to justify the evaluation of the lack of helping the person out of the car crash although it was physically impossible for A to do so.

We can differentiate between consequentialist and deontological evaluations strategies. In both these cases, we need to know more about the soft constraints that might impede or advance either the internal or external environment within which A acts. This supports the argument that the account of morally relevant can-conditions in ought-judgements is not merely a question of description and causality but requires normative criteria that allow for determination of which agents of those who can should do it. I return to this issue in the last section of this chapter.

Let us now turn to the soft constraints. Here things get a bit blurry. The reason for this, as Gilabert and Lawford-Smith have argued, is that whilst hard constraints are *binary*, which means they either make action X feasible or not, soft constraints are *scalar*, which means that they can make action X more or less feasible or more or less likely (Gilabert and Lawford-Smith 2012, 814–815). Add to this the possibility that soft constraints may change for various reasons, as we saw earlier. In order to adapt to this situation, some have suggested extending the concept of moral duties to what they call 'dynamic duties', which captures the idea that people's set of duties might have a dynamic property because they change over time.[3]

Here, we arrive at one of the main challenges of this inquiry. How do we address this type of changeable soft constraints on ought-judgements in normative reasoning about ought-judgements? The challenge resembles some of the problems related to the discussion of contextualism and the spilt-level theory of global justice discussed in the previous chapter. Consider this example. The nation-state has for the last 100–150 years been a politically significant institution in global society. This might change, and possibly is already changing, with the growing importance of economic globalisation and cooperation with non-state actors, such as non-governmental organisations (NGOs), international governmental organisations (IGOs), transnational corporations, and citizens involved in global policymaking (Goldin 2013). In this context, we may say that the new transnational institutions establish new soft external constraints that either impede or stimulate the feasibility or likelihood of certain acts and policies. Therefore, the normative examination of global justice cannot remain ignorant of what types of institutions and practices are governing the world.

Normally, we do not consider these types of soft external constraints as morally significant. As we have already seen, these constraints have no impact on how we understand what justice is or what persons are morally entitled to. However, if the argument for fact-sensitive political theory is sustained, we may nonetheless attribute moral significance to these kinds of soft external constraints. We may do so not in the sense that they are determinant of fundamental moral principles, but in the sense that the soft external constraints play a role in the determination of what concrete ought-judgements ought to be assigned to particular agents. If we accept this understanding of ought-assignments, the absence of soft external constraints, which make certain actions unlikely and unfeasible, is a necessary condition for the assignment of ought-judgements to particular agents. Again, it should be noted that they are not a sufficient condition. But in order to assign moral responsibility and ought-judgements to particular agents, the soft constraints are necessary.

To summarise, hard constraints are necessary but not sufficient conditions for determining who is a relevant moral agent for particular ought-assignments. Soft constraints are necessary but not sufficient conditions for fitness-conditions for which agents are fit to be held responsible for X. Together, the hard and soft conditions are sufficient can-conditions. The next section continues the scrutiny of the difference between internal and external soft constraints.

Soft constraints

We need to further substantiate what soft internal constraints we find valid in the discussion of moral responsibility. One group of criteria concerns intentional ability, which implies constraints on intentions, will, and trying (Mason 2003, 322–325). Many have rejected intentional constraints as legitimate conditions for what the agent is capable of doing. In particular, those who reject the OIC principle take the OIC to imply that can-conditions are identical to 'intentional ability' (Sinnott-Armstrong 1984, 252–254). The reason why intentional ability cannot be accepted as sufficient morally relevant can-conditions for ought-judgements is that it seems implausible to argue that people are not obliged to do X just because they do not *want* to or that people are not responsible for X just because they cannot do X *now*.

These two examples are vulnerable to what Frances Howard-Snyder called 'self-imposed inability' (Howard-Snyder 2006, 235). Consider the following example that Howard-Snyder provided: Alice promises to pay the money on Sunday and then spends it all on Saturday. On Sunday, she cannot pay the money because she has spent it all on Saturday.

For Sinnott-Armstrong, this provides justification for arguing that OIC only has what he calls 'conversationally meaning' (Sinnott-Armstrong 1984, 255).[4] Others, however, use this type of cases to illustrate that it would defeat the purpose of assigning ought if the agent cannot do it. The proponents of OIC will not blame Alice for the fact that she *cannot* pay the money Sunday. What they want to hold her responsible for is her actions on Saturday that made her unable to pay the money on Sunday.

Consider the following wedding case. Louis and Sarah are getting married in Boston at 9 am. At 8:30 a.m., Louis checks in for a flight to Los Angeles. At 8:57 a.m. he calls Sarah to say that he is on a plane to Los Angeles. The natural response for Sarah would be to say: 'But you *ought* to be here now' (Vranas 2007, 178). Opponents of the OIC would clearly say that Louis still has an obligation to be in Boston at 9 a.m. Proponents of the OIC, on the other hand, take a different view. They argue that at 8.57 a.m. Louis is no longer obliged to show up at 9 a.m. in Boston because he cannot. According to this understanding, what the utterance *you ought to be here now* most plausibly implies is that Louis ought not to have boarded the plane to Los Angeles at 8.30 a.m. So what Louis is blameworthy for is not that he did not show up at 9 a.m., but that he boarded a plane which made it impossible for him to show up in Boston at 9 a.m.

The reason for this is that ought-statements, which include can-conditions and questions of ability, have, as Gilabert and Lawford-Smith have argued, a 'diachronic' aspect which is related to the idea of dynamic duties. To what extent people can bring a state of affairs about is not merely a question of what they can do now at T, but also whether they have had the ability to '*get themselves into a position to be able* to bring it about' (Gilabert and Lawford-Smith 2012, 811). So conceived, the OIC principle cannot be rejected merely because Alice has a self-imposed inability on Sunday to pay the money. What is relevant in ought-judgements is whether Alice has tried to bring about a situation on Sunday where she is capable of paying the money.

Along the same lines, Howard-Snyder argued that in order to make one capable of doing X at T, there are several 'necessary enablers' that the agent ought to provide (Howard-Snyder 2006, 239). In other words, in order to avoid the self-imposed inability critique, soft internal constraints do not merely concern people's will, intentions, and ability at T, but whether they have *tried* to put themselves into a position at T that enables the agent to be able to do X at T. Elinor Mason defined 'trying' as a sub-class of intentional action: 'An agent tries to do X when she believes that X is logically and physically possible, and she does what she believes is most likely to result in X being brought about' (Mason 2003, 323).

Despite the fact that soft internal conditions such as intentional ability, trying, and will might account as necessary conditions for ought-judgements, they cannot provide a sufficient condition. One reason for this is that in many cases agents, at least in the case of climate change, which we shall see later, do not control the necessary enablers and the dynamic aspects of their own ability to do X at T (Howard-Snyder 2006, 236). Although people might 'try' to put themselves in a situation that enables them to do X at T, they will not succeed if they, for example, do not know what the best available evidence would suggest them to do (Mason 2003, 328; Gilabert and Lawford-Smith 2012, 823).

Moreover, we should be concerned about whether the social context, or what I later call the *behavioural space* within which agents act, provides the necessary external enabling conditions for the agents. Appreciating this type of soft *external* constraints provides a different account of what are the necessary and sufficient can-conditions with regard to the assignment of moral responsibility. It is important to draw attention to the soft external conditions under which agents are fit to be held responsible. These types of conditions for responsibility can be fleshed out in different ways.[5] One account determines three conditions that must be satisfied for someone to be fit to be held responsible. List and Pettit have established them in the following way (2011, 155):

1   *Normative significance*: The agent faces a normatively significant choice involving the possibility of doing something good or bad, right or wrong.
2   *Judgemental capacity*: The agent has access to, and understanding of, the evidence required for making normative judgements about the options.
3   *Relevant control*: The agent has the control required for choosing between the options.

These three conditions are external in the sense that they are dependent on the social, institutional, and political constraints available within the agent's behavioural space. The three conditions are individually necessary and sufficient to determine fitness to be held responsible (List and Pettit 2011, 155). In order to be fit to be held responsible, the agent should face a normatively significant *choice* over which she has *knowledge* and *control*.[6] This account does not necessarily neglect the importance of intentions, trial, and will, but they are not the primary properties considered relevant as can-conditions for ought-judgements. However, depending on which methodology is assumed, the internal and external soft constraints will be weighed differently. As argued, the discussion of *can-* and *fitness-* conditions is deeply embedded in what methodological account of agency and rationality is presumed and accordingly, how the social and political problems of climate change are perceived.

Methodological assumptions

Let us consider one climate-related example that shows what methodological assumptions might be considered relevant. In the international climate politics and climate ethics literature, the 'polluter pays principles' (PPP) dominates.[7] The PPP holds those who cause climate change responsible: those who cause X are morally responsible for causing X and for paying the cost of remedying X.[8] This approach to individual responsibility relates the question of responsibility to economic questions of costs and benefits distribution – that is, who should bear the costs of adaptation or mitigation policies and who should compensate victims harmed by climate change. The economic framing of the climate debate has been importantly influenced by the report *The Stern Review* (N. Stern 2006).[9] Likewise, climate ethical research often takes its point of departure from discussions of economic cost-benefit analysis and the challenges related to the *prisoner's dilemma* and *tragedy of the commons* (Sagoff 2008; Parfit 2010; Gardiner 2011; Broome 2012). The theoretical framework is associated with methodological individualism and rational choice theory.

Although many climate ethics scholars criticise the arguments and normative implications of the self-interest and rational choice theory, to a large extent, the debate still takes its point of departure from methodological individualism.[10] Methodological individualism argues that the level of individual agency is the primary explanation of the individual's choices and acts. As we shall see in Part 2, these assumptions have implications for how to perceive the fitness-conditions that must be present for the agent to be held morally responsible for climate change. The agents' can- and fitness-conditions are centred around the cost and benefits trade-off that is thought to motivate the rational agents' actions.

It is standard in the political and social sciences to distinguish individualist, collectivist, and institutionalist methodologies.[11] Individualism can be fleshed out by reference to the individual agent's *psychology* (behaviourism), *rationality* (cognitivism), or *emotions* (emotivism) (Cox 2004). The rational-individualist methodology attributes neither moral autonomy nor explanatory power to the group and institutional levels of agency. Rational-individualism claims that 'anything ascribed to a group [. . .] can be re-expressed by reference to its members' (Coleman 1994, 2; List and Pettit 2011, 3). To the individual-rationalist, a government, for example, is merely a collection of individually acting singular persons, not a collective entity.[12] Modified theories of individual-rationalism, which we consider in more detail later, include some social and behavioural norms that aim to integrate the level of institutions into the methodology of individualism (Ostrom 1990). For reasons that will become apparent later, this modification does not change much because the rational choice assumptions are maintained.

On the other hand, collectivists claim that anything ascribed to a singular individual can be re-expressed by reference, for example, to its class, culture, societal structure, or level of technological development. Proponents of collectivist methodologies reject individualism by explaining social and political phenomena by referring to social phenomena such as culture, class, history, and technology (Heath 2011; List and Pettit 2011). The individualist and collectivist methodologies take an *eliminative* approach to explanations and rational reasoning about agency, reducing the explanations of social events and individual acts to one level of agency.

In contrast to what we could call *single-level* methodologies of individualism and collectivism, the supervenience theory, which will be elaborated upon later, allows for a *non-reductionist* methodology. If we accept the supervenience theory, the individual and collective levels of agency are non-reducible to each other but are at the same time reciprocal. The non-reductionist model allows for both individualist and institutionalist methodologies to be appreciated. This non-reductionist methodology of institutional group agency allows for lower and higher levels of agency and rationality to co-exist.

Therefore, the non-reductionist methodology of institutionalism rejects the rationalistic version of methodological individualism in which individuals are perceived merely as sullied rational agents (Shue 1988; Hurley 2011, 202–209; List and Pettit 2011, Chapter 3).[13] Instead, the assumption is that individual agents form part of a complex network of social norms, multiple functions, and multifaceted institutions. The institutional and behavioural spaces influence the set of norms and incentives of the individual agents. This non-reductionist methodology also provides fruitful ground for group agency and, as we shall see later, for discussing assignment of responsibility for climate change to institutionalised group agents. Moreover, the non-reductionist framework allows for the assignment of explanatory power to social norms, institutional designs, and societal functions without relinquishing the level of individual agency. Depending on the institutional context, level of division of labour, and integrated intentionality, different explanatory models will prevail (Le Grand 2006; List and Pettit 2011; Hurley 2011).

Group agency

To complete the account of which can- and fitness-conditions are relevant for climate responsibility, let us consider the concept of group agency. We shall see that some group agents are fit to be held collectively responsible for climate change. Thus, by invoking the non-reductionist framework, it is not merely the explanatory power that is shifted but also the basic anthropological assumptions about human agency and arguably, the agents' can- and fitness-conditions to be held morally responsible. This methodology allows for not merely individual agents to be possible moral agents that are blameworthy for climate change; institutionalised group agents also may be blameworthy agents if they can do what ought to be done and moreover satisfy the normative criteria for responsibility. More particularly, Part 2 discusses whether the theories succeed or fail to satisfy the three criteria.

Theories of individual responsibility that presume rationalistic assumptions of self-interest and short-sightedness fail to satisfy the conditions of normatively significant choice and access to knowledge. The primary reason for this failure is that individual agents are expected to neglect the collective benefits of environmental sustainability. Even if we do not take a narrow rationalistic approach to individual agency by incorporating ideas of social norms and institutions, individuals do not have control over morally significant choices in relation to climate change. In contrast, I argue that democratic institutions are better capable of providing necessary external enablers that allow group agents to meet the three criteria and thereby prove fit to be held collectively responsible for climate change.

There are different ways of conceptualising group agency. Let us begin by outlining one simple distinction between loosely connected group agents and institutionalised group agents. Loosely connected group agents are uncoordinated and arbitrary groups, such as a group waiting for the bus,[14] whereas coordinated and formalised institutions are, for example, a governmental body, legislative agency, and other types of public and private institutional bodies at the subnational, national, and supranational levels (List and Pettit 2011, Chapter 7). For present purposes, I focus on the coordinated group agency attributed to formal institutions in Western democracies. I focus in particular on the social institutions of the market and political democracies.[15]

In contemporary political science and political philosophy, there is no generally agreed-upon theory of agency remains controversial. It is possible to agree on the basic fact that there exist formal institutions in modern democracies but nonetheless attribute no explanatory power to them and thus consider them unfit to be held collectively responsible. This is due to the different methodological assumptions mentioned earlier. Recent contributions exhibit a new interest in institutionalism that addresses the social and institutionally mediated context within which individual agents act and deliberate (Ostrom 2005; Searle 2010; List and Pettit 2011). These institutionalist approaches draw attention to the level of political organisation, governance, and agency that is irreducible to either micro approaches (rational single policymakers) or macro-approaches (societal forces such as technological innovation, cross-cultural migration, and demographic change) (Olsen 2007, 3). They analyse the nature, status, and significance of institutional functions, institutionalised collective groups, and generally the institutional complexity of political organisations as agents and bodies in their own right.

It is common in the field of political science to recognise an input side and an output side of institutions (Easton 1965; Bang 2008).[16] The complexity of the institutionalised group agency originates from the fact that the input side and output side are not directly connected. The input side encompasses the motives and intentions that may initiate various collective and institutional processes, enterprises and policies. Both cultural 'we-intentions' and individual and collective intentionality play a role here (Tuomela 1991, 2007; Tomasello et al. 2005; Searle 2010). On the other hand, output conditions consist of the policy implications of the initiated processes. The total outcome of the collective process cannot be reduced to the initial decision, motive, or intention of the input side. Institutional group agency facilitates multilevel and complex governance networks in which decisions, motives, and intentions are negotiated, transformed, and influenced by agents at all levels (Bang and Esmark 2009). In Part 3, I demonstrate how this complex structure of institutionalised group agents has implications for what we may hold them responsible for.

Nonetheless, many scholars have been critical of the theory of group and institutional agency (Popper 2002; Hayek 1955). They criticise the assumption that group agents are presumed to constitute an independent ontological level of reality. However, there are two traditions of thinking about group agents. One is the tradition of *emergentism*, in which the collective level of agency expresses a transcendent metaphysical reality. Another is the tradition of *supervenience*, in which the collective agent has a robust autonomous level of agency that nonetheless supervenes on its single constituencies. According to the latter, institutions have no ontologically independent reality (List and Pettit 2011, 73–76). Both traditions remain contested. Emergentism is a more demanding position because the supervenience thesis allows for a level of group agency without collapsing the individual level of agency (List and Pettit 2011, 73–76).

The basic idea of supervenience is that the collective and individual levels of agency are connected but are nonetheless two autonomous levels of agency. The group level of agency supervenes on individual agents, but the individual level of agency remains incompletely determined and explained by the group agent. In the following discussion, I take the supervenience thesis to be a plausible account of group agency because it allows for a level of collective agency without presuming a demanding metaphysical theory of group agency. In order to put more flesh on the supervenience thesis, let us consider what constitutes the supervenience relationship between the individual and collective levels of agency. Some have suggested that the relationship consists of the group agent's organisational structure, design, and functionality:

> Given that talk of group agents is not readily translatable into individualistic terms, and given that it supports a distinct way of understanding and relating to the social world, we can think of such entities as autonomous realities. Although their agency depends on the organization and behaviour of individual members, as individualism requires, they display patterns of collective behaviour that will be lost on us if we keep our gaze fixes at the individual level. And to lose sight of those patterns is to lose an important source of guidance as participants in the social world.
>
> (List and Pettit 2011, 6)

This means that the existence of the group agent depends not only on the single individuals but also in the way they coordinate and cooperate their activities. Examples of such structures are the social institutions of the marketplace and democracy. Put differently, the supervenience relationship is determined by a complex network of reciprocally constituting activities, which allows different levels of agency to supervene on each other. So conceived, the multi-level and complex structure of group agency resembles the relationship between the brain and its neurons. In this analogy, the infrastructure of the group agent is the neuronal pathways that together constitute an institutionally and socially nested grid. This enables us to appreciate 'higher-level regularities that can guide us in thinking about the outputs likely to result from various inputs, and in determining how we should respond to the system in question, often in abstraction from much of what happens at the lower level' (List and Pettit 2011, 6).

According to this argument, the group agent supervenes on the constantly developing and changing network of individual acts (List and Pettit 2011, 77–78), or what Miller calls 'a multi-layered structure of joint actions' (S. Miller 2011, 223). When we accept this approach, which invokes ideas of supervenience and institutionally mediated group agency, we have reason to investigate whether these institutionalised group agents are morally responsible agents. I show that they ought to be held more responsible for rectifying the negative effects of climate change than individual agents are.

Moral premises

At this juncture, two related normative conditions of the OIC principle should be outlined. The normative conditions allow for the ought-assignment to be given to particular agents that not merely *can* but also *should* do what ought to be done. It is important to stress that the point of departure of the discussions is not whether a fundamental moral principle of, for example, climate ethics, applies or not; rather, that the point of departure is the moral claim that something ought to be done, that is, climate change needs to be remedied. Until now we have been preoccupied with the role of the can- and fitness-conditions in ought-assignments, which can be summed up in this way:

OIC#1

A ought to do X

1 If and only if there is a moral wrongdoing;
2 If and only if A is a moral agent; and
3 If and only if A can do X.

Scholars have argued that the can- and fitness-conditions are necessary premises for a normative theory of assigning who ought to do something about climate change. But they provide an insufficient account of who should be assigned responsibility to specific agents. In order to assign responsibility to particular agents, it is not enough that the agents can do it: two further normative criteria need to be invoked. First, it may be the case that an agent can do something but for various reasons can be morally excused:

4 If and only if A is not morally excused

To excuse someone morally presumes two things. First, the blameworthy act should have been avoided or 'where there has been some abnormality or failure' (Austin 1968, 23). Second, the blameworthy person is morally obliged to avoid the act (under normal conditions). In the context of climate change, if someone is morally excused, it implies that the act should have been avoided and that the excused person has moral obligations towards, for example, the earth's atmosphere and climate and future generations.[17] In Part 3, I examine two reasons to excuse moral agents for contributing to climate change. One is an *intentionalist* reason, which links blameworthiness of an act with the intentions the person had whilst pursuing the act. If an agent breaks the speed limit because the purpose (i.e., her intentions) is to get a critically ill passenger to the hospital, we would probably not blame her for driving too fast. And one is a *consequentialist* reason, which links blameworthiness to the total outcome of the act(s). As we shall see, the excuses that apply at the individual level of agency do not apply the collective level of agency. This is because of the account of methodology and group agency defended earlier.

Second, there needs to be a positive justification of why it may be fair to assign responsibility to particular agents. Good candidates for justification could be efficiency or democratic legitimacy. Thus, the fourth premise of my account of ought-assignments is:

5 If and only if there is a positive justification of why A should be held morally responsible

One reason why we need a positive justification is that if one accepts that soft constraints do not merely concern intentions – will and try, but also choice, knowledge, and control – the implication is that the theory of ought-judgement becomes more context- or policy-specific. If choice, knowledge, and control are significant can-conditions in ought-judgements, the account of ought-judgements will necessarily differ from situation to situation. Thus, an agent's responsibility will most likely change according to the topic in question. It might be that in the case of environmental harm some agents have sufficient choice, knowledge, and control and are therefore fit to be held responsible, whilst the agents in the situation of climate change do not have sufficient choice, knowledge, and control and are therefore not fit to be held responsible. In Part 2, I continue the discussion of responsibility and argue that the fit-conditions of responsibility play a normatively significant role when assigning responsibility for the policy-specific case of climate change.

Conclusion

I have focused in this chapter on a notion on ought-assignment that is fundamental to the understanding of moral responsibility for climate change. I have claimed and defended the view that responsibility understood as a way of assigning ought-judgements to particular agents relies on specific criteria for what the agent *can* do. An important discussion in this respect was to defend the view that it is possible to accept the premise that concrete ought-assignments are contingent on methodological assumptions about human agency and capacities without neglecting the existence of fundamental normative principles. I have also outlined an account of ought-assignments that includes premises about moral excuse and positive justification, to which we will return in Part 3.

Notes

1 A different way to understand this argument could be that the abstract principles are conditionals, which would make the distinction redundant. For the present purposes, I set this option aside.

2 For contributions on the dynamic aspects of institutions, see, e.g., Shue (1980), Rawls (1993), Sangiovanni (2008), Ostrom (2005), Sen (2009), and Gilabert (2012).

3 For example, Gilabert argued for integrating a 'dynamic aspect' into political theory: 'Prescriptions of justice operate at different levels, with different feasibility constraints. Principles involve more general constraints than their implementation, and the latter involve more general constraints than the strategies of reform leading to them. These differences are important to develop a dynamic approach to facts blocking the pursuit of justice. We can imagine alternative implementations of principles, and we may have dynamic duties to reshape feasibility constraints over time. When we do this, our deliberations about how to pursue justice take a long-term transitional standpoint in which we are both normatively ambitious and hard-nosed about the realities of social life' (Gilabert 2012, 25).

4 'Roughly, saying *p* conversationally implies *q* when saying *p* for a certain purpose cannot be explained except by supposing that the speaker thinks that *q* and thinks that the hearer can figure out that the speaker thinks that *q*, etc.' (Sinnott-Armstrong 1984, 256).

5 For an overview of different positions, see Jamieson (2010). For other approaches to moral responsibility, see Scanlon (1998) and Korsgaard (1996).

6 The discussion of control touches upon the philosophical debate on determinism and free will (e.g., Fischer and Ravizza 1998). In the present context, control refers to whether relevant choices are available within what later will be called institutional behavioural spaces.

7 See, e.g., Shue (1999), Singer (2002), Gosseries (2004), Caney (2005b), Page (2006), Miller (2008), and Jagers and Duus-Otterström (2008). Notably, the Organization for Economic Cooperation and Development (OECD) has adopted and recommended PPP (1992). Furthermore, the European Union and the Council of Ministers have affirmed PPP (Directive 2004/35/CE; Caney 2010a, 124).

8 Alternative theories are the *polluter pays principle* (PPP), the *beneficiary pays principle* (BPP) and the *ability to pay principle* (APP), which assign responsibility to agents independently of causal responsibility. Instead, those who have benefited from X are morally responsible or those who are capable of paying for X are morally responsible to do so – even though neither group caused X. For a discussion of several different accounts of PPP, BPP, and APP, see Caney (2010).

9 *The Stern Review on the Economics of Climate Change* was produced for the British government and published on 30 October 2006 by economist Nicholas Stern.

10 This is also the point of departure in discussions on luck-egalitarianism (Arneson 2010; Cohen 2011) and distributive justice (Rawls 1971; Dworkin 2000). For recent contributions, see Knight and Stemplowska (2011). For general discussion, see, e.g., Cohen (1997) and Hurley (2011). For discussions with regard to climate ethics, see Sagoff (2008), Caney (2010a), Gardiner (2010), and Shue (2010).

11 For methodological individualism, see Popper (2002), Hayek (1955), Elster (1982b), and Coleman (1994). For classic methodological collectivism, see Durkheim (1964). A version of methodological collectivism is also frequently attributed to scholars such as Auguste Comte and Karl Marx (Heath 2011). For methodological institutionalism, see Ostrom (1990, 2005). For a mixed approach, see, for instance, rational choice institutionalism (Hall and Taylor 1996).

12 Margaret Thatcher is famous for once having said, 'There is no such thing as society' (Thatcher 1987). In this context, every act and norm of the government can be completely explained by referring to each member of the group. Only individuals exist.

13 The communitarian critique is raised for the same reasons but stresses the cultural, social, and national contexts instead of the institutional context. See, e.g., Kymlicka (1989, 1995) and Miller (1995).

14 See, e.g., Hardin (1988), Held (2002), and S. Miller and Makela (2005). For a recent discussion, see Lippert-Rasmussen (2011, 104–114).

15 Other relevant group agents could be private corporations, NGOs, and IGOs. Despite their importance to contemporary global climate politics, I set these aside in this discussion.

16 For a critical approach to the input/output distinction, see Scharpf (1973, Chapter 1).

17 See, e.g., Caney (2005a) for a good overview.

References

Arneson, R. J., 2010. Self-ownership and world ownership: against left-libertarianism. *Social Philosophy and Policy*, 27 (Winter), 168–194.

Austin, J. L., 1968. A plea for excuses. *In:* A. R. White, ed. *The philosophy of action.* Oxford: Oxford University Press, 19–42.

Bang, H., 2008. Between democracy and good governance. *JPoX – Journal on Political Excellence* [online]. Available from: http://jpox.eu/component/streams/view.content/cid.211/ [Accessed 12 May 2016].

Bang, H. and Esmark, A., 2009. Good governance in control society: reconfiguring the political from politics to policy. *Administrative Theory and Praxis*, 31 (1), 7–37.

Björnsson, G., 2011. Joint responsibility without individual control: applying the explanation hypothesis. *In:* N. Vincent, I. van de Poel, and J. van den Hoven, eds. *Moral responsibility: beyond free will and determinism*. Dordrecht: Springer, 181–200.

Broome, J., 2012. *Climate matters: ethics in a warming world*. London: W.W. Norton.

Caney, S., 2005a. *Justice beyond borders: a global political theory*. Oxford: Oxford University Press.

Caney, S., 2005b. Cosmopolitan justice, responsibility, and global climate change. *Leiden Journal of International Law*, 18 (4), 747–775.

Caney, S., 2010. Climate change, human rights and moral thresholds. *In:* S. Humphreys, ed. *Human rights and climate change*. New York, NY: Cambridge University Press, 69–90. Reprinted in 2010: *In:* S. M. Gardiner, S. Caney, D. Jamieson, and H. Shue, eds. *Climate ethics: essential readings*. Oxford: Oxford University Press, 163–180.

Cohen, G. A., 1997. Where the action is: on the site of distributive justice. *Philosophy and Public Affairs*, 26 (1) (Winter), 3–20.

Cohen, G. A., 2008. *Rescuing justice and equality*. Oxford: Oxford University Press.

Cohen, G. A., 2011. *On the currency of egalitarian justice, and other essays in political philosophy*. Princeton, NJ: Princeton University Press.

Coleman, J., 1994. *Foundations of social theory*. Cambridge, MA: Harvard University Press.

Cox, G. W., 2004. Lies, damned lies, and rational choice analyses. *In:* I. Shapiro, R. M. Smith, and T. E. Masoud, eds. *Problems and methods in the study of politics*. Cambridge: Cambridge University Press, 167–185.

Durkheim, E., 1964. *The rules of sociological method*. New York, NY: The Free Press.

Dworkin, R., 2000. *Sovereign virtue*. Cambridge, MA: Harvard University Press.

Easton, D., 1965. *A system analysis of political life*. New York, NY: Wiley.

Elster, J., 1982b. The case for methodological individualism. *Theory and Society*, 11 (4), 453–482.

EU Commission, 2004. *Directive 2004/35/CE of the European Parliament and of the council on environmental liability with regard to the prevention and remedying of environmental damage*. Available from: http://europa.eu.int/eur-lex/pri/en/oj/dat/2004/1143/1_143 20040430 en00560075.pdf [Accessed 3 April 2016].

Fischer, J. M. and Ravizza, M., 1998. *Responsibility and control: a theory of moral responsibility*. Cambridge: Cambridge University Press.

Frankfurt, H., 1969. Alternate possibilities and moral responsibility. *Journal of Philosophy*, 66 (23), 829–839.

Gardiner, S. M., 2010. Ethics and global climate change. *In:* S. M. Gardiner, S. Caney, D. Jamieson, and H. Shue, eds. *Climate ethics: essential readings*. Oxford: Oxford University Press, 3–38.

Gardiner, S. M., 2011. *A perfect moral storm: the ethical tragedy of climate change*. Environmental Ethics and Science Policy Series. New York, NY: Oxford University Press.

Gilabert, P., 2012. *Dynamic feasibility, principles of justice, and all-things-considered political judgment*. Presented at the conference Facts and Norms, University of Copenhagen, 22–23 August 2012.

Gilabert, P. and Lawford-Smith, H., 2012. Political feasibility: a conceptual exploration. *Political Studies*, 60 (4), 809–825.

Goldin, I., 2013. *Divided nations: why global governance is failing, and what we can do about it*. Oxford: Oxford University Press.

Gosseries, A., 2004. Historical emissions and free-riding. *Ethical Perspectives*, 11 (1), 36–60.

Hall, P. A. and Taylor, R., 1996. Political science and the three institutionalisms. *Political Studies*, 44, 936–957.

Hardin, R., 1988. *Morality within the limits of reason*. Chicago, IL: University of Chicago Press.

Hayek, F., 1955. *The counter-revolution of science*. New York, NY: Free Press.

Heath, J., 2011. Methodological individualism. *In:* E. N. Zalta, ed. *The Stanford Encyclopedia of Philosophy*. Spring edn. Available from: http://plato.stanford.edu/archives/spr2011/entries/methodological-individualism/ [Accessed 16 May 2016].

Held, V., 2002. Group responsibility for ethnic conflict. *Journal of Ethics*, 6 (2), 157–178.

Howard-Snyder, F., 2006. 'Cannot' implies 'not ought'. *Philosophical Studies*, 130, 233–246.

Hurley, S., 2011. The public ecology of responsibility. *In:* C. Knight and Z. Stemplowska, eds. *Responsibility and distributive justice*. Oxford: Oxford University Press, 187–215.

Jagers, S. C. and Duus-Otterström, G., 2008. Dual climate change responsibility: on moral divergences between mitigation and adaptation. *Environmental Politics*, 17 (4), 576–591.

Jamieson, D., 2010. Ethics, public policy, and global warming. *In:* S. M. Gardiner, S. Caney, D. Jamieson, and H. Shue, eds. *Climate ethics: essential readings*. Oxford: Oxford University Press, 77–86.

Knight, C. and Stemplowska, Z., eds., 2011. *Responsibility and distributive justice*. Oxford: Oxford University Press.

Korsgaard, C. M., 1996. *The sources of normativity*. Cambridge: Cambridge University Press.

Kymlicka, W., 1989. *Liberalism, community and culture*. Oxford: Clarendon Press.

Kymlicka, W., 1995. *Multicultural citizenship*. Oxford: Clarendon Press.

Le Grand, J., 2006. *Motivation, agency, and public policy: of knights and knaves, pawns and queens*. New York, NY: Oxford University Press.

Lippert-Rasmussen, K., 2005b. Determinism, compatibilism, and incompatibilism. *In:* K. Lippert-Rasmussen, ed. *Deontology, responsibility, and equality*. Copenhagen: Copenhagen University Press, 46–47.

Lippert-Rasmussen, K., 2011. We are all different: statistical discrimination and the right to be treated as an individual. *Journal of Ethics*, 15 (1–2), 47–59.

List, C. and Pettit, P., 2011. *Group agency: the possibility, design, and status of corporate agents*. Oxford: Oxford University Press.

Mason, E., 2003. Consequentialism and the 'ought implies can' principle. *American Philosophical Quarterly*, 40 (4), 319–331.

Miller, D., 1995. *On nationality*. Oxford: Oxford University Press.

Miller, D., 2008. Political philosophy for earthlings. *In:* D. Leopold and M. Stears, eds. *Political theory: methods and approaches*. Oxford: Oxford University Press, 29–48.

Miller, S., 2010. *The moral foundations of social institutions: a philosophical study*. Cambridge: Cambridge University Press.

Miller, S., 2011. Collective responsibility, epistemic action and climate change. *In:* N. Vincent, I. van de Poel, and J. van den Hoven, eds. *Moral responsibility: beyond free will and determinism*. Dordrecht: Springer, 219–246.

Miller, S. and Makela, P., 2005. The collectivist approach to collective moral responsibility. *Metaphilosophy*, 36 (5), 634–651.

OECD (Organization for Economic Co-operation and Development), 1992. *General Distribution OCDE/GD (92) 81. The polluter-pays principle: OECD analyses and recommendations*. Paris: OECD.

Olsen, J., 2007. *Europe in search of political order: an institutional perspective on unity/diversity, citizens/their helpers, democratic design/historical drift and the co-existence of orders*. Oxford: Oxford University Press.

Ostrom, E., 1990. *Governing the commons: the evolution of institutions for collective action*. Cambridge: Cambridge University Press.

Ostrom, E., 2005. *Understanding institutional diversity*. Princeton, NJ: Princeton University Press.

Page, E., 2006. *Climate change, justice and future gener*ations. Cheltenham: Edward Elgar.

Persson, I. and Savulescu, J., 2012. *Unfit for the future: the need for moral enhancement*. Oxford: Oxford University Press.

Popper, K., 2002. *The poverty of historicism*. Berkshire: Routledge. *In:* German: Popper, K., 1957. *Das elend des historizismus*. Tübingen: Mohr.

Rawls, J., 1971. *A theory of justice*. Cambridge, MA: The Belknap Press of Harvard University Press.

Rawls, J., 1993. *Political liberalism*. Cambridge, MA: Harvard University Press.

Sagoff, M., 2008. *The economy of the earth, philosophy, law and the environment*. New York, NY: Cambridge University Press.

Sangiovanni, A., 2008. Justice and the priority of politics to morality. *The Journal of Political Philosophy*, 16 (2), 137–164.

Scanlon, T. M., 1998. *What we owe to each other*. Cambridge, MA: Harvard University Press.

Scharpf, F., 1973. [Planning as a political process: essays on the theory of planning democracy] *Planung als politischer prozess: aufsätze zur theorie der planenden demokratie*. Frankfurt Am Main: Suhrkamp Verlag.

Searle, J., 2010. *Making the social world: the structure of human civilization*. New York, NY: Oxford University Press.

Sen, A., 2009. *The idea of justice*. London: Penguin Books.

Shue, H., 1988. Mediating duties. *Ethics*, 98 (4), 687–704.

Shue, H., 1980. *Basic rights: subsistence, affluence, and US foreign policy*. Princeton, NJ: Princeton University Press.

Shue, H., 1999. Global environmental and international inequality. *International Affairs*, 75 (3), 531–545. Reprinted in 2010. *In:* S. M. Gardiner, S. Caney, D. Jamieson, and H. Shue, eds. *Climate ethics: essential readings*. Oxford: Oxford University Press, 101–111.

Singer, P., 2002. *One World*. New Haven, CT: Yale University Press.

Sinnott-Armstrong, W., 1984. 'ought' conversationally implies 'can'. *The Philosophical Review*, 93 (2), 249–261.

Stern, N., 2006. What is the economics of climate change? *World Economics*, 7 (2), 1–10.

Stern, R., 2004. Does 'ought' imply 'can'? And did Kant think it does? *Utilitas*, 16 (1), 42–61.

Thatcher, M., 1987. *Interview for Woman's Own ('no such thing as society')*, 23 September. Available from: www.margaretthatcher.org/document/106689 [Accessed 5 September 2016].

Tomasello, M., Carpenter, M., Call, J., Behne, T., and Moll, H., 2005. Understanding and sharing intentions: the origins of cultural cognition. *Behavioral and Brain Sciences*, 28 (5), 675–735.

Tuomela, R., 1991. We will do it: an analysis of group-intentions. *Philosophy and Phenomenological Research*, 51 (2), 249–277.

Tuomela, R., 2007. *The philosophy of sociality: the shared point of view*. New York, NY: Oxford University Press.

Vincent, N. and van de Poel, I., 2011. Introduction. *In:* N. Vincent, I. van de Poel, and J. van den Hoven, eds. *Moral responsibility: beyond free will and determinism.* Dordrecht: Springer, 1–14.

Vranas, P. B. M., 2007. I ought, therefore I can. *Philosophical Studies*, 136 (2), 167–216.

Part 2

Fitness-conditions of moral responsibility

4 Fitness-conditions of rational agency

In my account of fact-sensitive ought-assignments, moral responsibility cannot be reduced to moral blameworthiness. It depends also on a broader definition, including the can- and fitness-conditions, specifically the soft and hard constraints discussed in Part 1. In this Part, I am concerned only with soft constraints. I have suggested that a first defining property of soft constraints is *internal* conditions of intent or the goal to try to avoid climate harm, and that a second defining property is *external* conditions of significant choice, sufficient knowledge, and control. This investigation focuses on the soft external constraints, but the internal constraints will be addressed as well, particularly with regard to the individual agents. A third defining feature of my approach to the can- and fitness-conditions is the set of methodological assumptions that provide different understandings of who are the relevant moral agents.

First, I consider the fitness-conditions of individual agents as they are understood within the framework of rational-individualism. I consider whether this framework understands rational agency in a way that makes agents fit to be held responsible for the consequences of their choices. Second, I apply the earlier developed account of non-reductionist institutionalism and investigate whether certain types of institutionalised group agents may be fit to be held morally responsible for climate change. One point should be added: whilst rational-individualism assumes every agent to be rational, the non-reductionist institutionalist methodology allows for a relaxation of the rationalistic assumptions. To differentiate more easily between these two conceptualisations of agency, I differentiate *rational agents* from *democratic citizens*. In this context, rational agents are only motived by economic incentives and self-interest. In contrast, democratic citizens are also rational, but it is appreciated that they also have moral inclinations and other social properties. In other words, the rational agent's agent is rational, whereas democratic citizens are 'reasonable' (Rawls 1999; Elster 2009).

I argue that rational agents are unfit to be held morally responsible for climate change. This and other conclusions of this chapter fit well with other contemporary contributions that suggest alternative theories of responsibility for climate change. Particularly, I argue this point in the direction of an account of hybrid forms of responsibility and joint and collective responsibility, to which I return in Part 3 (Sinnott-Armstrong 2010; Vanderheiden 2011; S. Miller 2011).

Self-interest theory

In this chapter, I discuss the understanding of individual responsibility, agency, and rationality as it is understood in the individual rational methodology and investigate whether the rational-individualistic methodology allows for a normatively significant choice, sufficient knowledge, and control with regard to climate change at the individual level of agency. What interests us here is whether the agent is supposed to be free to choose to favour or disfavour climate-friendly behaviour. If individual-rationalism only proposes one set of choices (i.e., climate-unfriendly behaviour), rational agents do not face a normatively significant choice with regard to climate change.

More particularly, I discuss three variants of the rational choice situation. One choice situation is motivated by *self-interest*, another by *social norms*, and yet another by the *behavioural space* within which the choice happens. I look into each explanatory mechanism and explain why rational agents ostensibly act in an environmentally unfriendly manner. I then consider whether rational individuals have sufficient knowledge and control over the choice situation. I apply recent works of Elinor Ostrom, which draw attention to the role of social and institutional norms in climate ethics. I argue that the understanding of self-interest and cost and benefits calculations in the prisoner's dilemma games is too narrow a framework to comprehend the social, political, and moral challenges related to climate change (Broome 2012).

Let us begin by considering self-interest and rational choice theory, which understands agents' choices by reference to their 'beliefs to best satisfy their desires' (Cox 2004, 170). What we are interested in here is whether the theory of rational agency faces a normatively significant choice with regard to the future benefits and natural sustainability. The focus concentrates on the agents' rational goal-achievement since it is assumed that rational agents aim to obtain what benefits their individual interests most (Parfit 1984, 3; see also Williams 1973).[1] Self-interest theory argues that rational agents' short-term self-interest is equivalent to what is the overall aggregated public and long-term interest (Smith 2009, 158).

From this viewpoint, the phenomenon of climate change poses a paradox: there is no equivalence between individual and public interests. If we assume that a sustainable climate would be in the people's self-interest, the self-interest theory fails to explain why climate change is nonetheless the aggregated effect of the choices conducted by rational agents. Assuming that it is in people's long-term self-interest to have a sustainable environment and climate, people may believe that it is in their (short-term) self-interest not to avoid climate change, but in the long run, this procedure is self-defeating. By so arguing, the self-interest theories risk being indirectly self-defeating.[2]

To understand the logic of the self-interest theory, let us look at one assumption. The reason why individual rational strategies do not lead to collective optimal solutions is that rational agents rate the significance of the future to be less when compared to the present (Ostrom 1990). In economic theory, this is called the *discount rate*, which determines the net present value of benefits and costs to be received in the future. The assumption is that every rational agent will discount the value of the future to be less than the present. The discount rate is the mechanism with which economic-inspired theories can explain why it is rational for agents to rate the value of the future to be less when compared to values today. Intuitively, we would expect people to value their own future highly, but economic rational theory claims that rational agents value their present much higher than they do the future, and moreover, value personal values more highly than collective values (Parfit 1984, Chapter 1).

This implies that the agent's present interests may be inconsistent with the agent's interests as they turn out to be in the future. Consider the following example. One may have a lifelong self-interest in living in a sustainable environment. Currently, one enjoys participating in Formula 1 car racing. By participating in car racing, the agent does not act in the agent's long-term self-interest – that is, the present desire to participate in auto racing is inconsistent with a lifelong interest in a sustainable environment and climate.

On the other hand, it seems irrational to act in a completely time-neutral manner (Sidgwick 1981, 7–11) – i.e., to have a completely flat discount rate that values the present and the future equally. Because it is unreasonable (and practically impossible) to pay as much attention to one's life in, let us say, 50 years as to one's life at present, the self-interest and discount rate theories cannot explain how the rational agent balances both present and future interests. As such, there is an incompatibility between the notions of self-interests and discount rates as a way of valuing future values. Arguing that it is irrational not to pay attention to what happens in the future, Susan Hurley describes the economic-inspired self-interest theory as 'cognitive and behavioural anomalies' and questions the concept of rationality fundamentally (Hurley 2011, 195–202).

Moreover, Parfit criticises the economic and rational self-interest framework for predicting 'irrational rational' behaviour (Parfit 1984, Chapter 1). Individuals may have reasonable desires and aims but are expected to commit several estimation and deliberation errors with regard to collective benefits that lead to choices, acts, and values that run counter to the agents' long-term desires and values (Elster 1982; Parfit 2011, Chapter 3). Another problem of rational preference making is whether people's desires and values are adaptable to the opportunities that are available to them (Elster 1982, 219ff). If people's values were adaptable to the current available opportunities, this would provide another reason for why the theory does not provide a reliable account of what should be the rational agents' self-interest.

Another critical point should be added. It should be questioned whether it makes sense to discuss the moral significance of the future in economic discount rates. Why should the moral significance of the future of the planet and its people be dependent on an economic and rational theory of valuing?[3] Most commonly, discount rates differentiate events temporarily. However, as Parfit rightly points out, it is a mistake to discount values for pure time. The fact that negative consequences may be temporarily (or geographically) remote does not make them less likely or less important. Indeed, when applied further into the future, many predictions become more likely (Parfit 1984, 483). Alternatively, people could discount for probability. Again, as Parfit explains, it may be rational to be less concerned about the more remote effects of our current acts and behaviour, but this would never be because these effects are indeed more remote. Rather, it would be because they are less likely to occur (Parfit 1984, 486). In other words, if effects are likely to occur, it is rational to discount for them. Following this line of argument, we ought to be equally concerned about the predictable effects of our acts, whether these will occur 1, 100, or 1,000 years (Parfit 1984, 486).

To summarise this examination of the self-interest theory, we can say that the theory does not provide a strong framework in which rational agents face a normatively significant choice with regard to climate change. The theory predicts *irrational rational* behaviour that does not resemble the properties of a normatively significant choice. Note that in the non-reductionist methodology, which I discuss later, agents are also assumed to be rational; nonetheless, the economic rationality does not exhaust the intentions, motives, and desires of the agents.

Social dilemma theory

Let us see if the social dilemma theory is more successful in making rational agents fit to be held responsible for climate change. The challenge of climate change resembles what Garrett Hardin in his seminal article in *Science* (1968) has called *the tragedy of the commons.*[4] The *tragedy of the commons* challenges the self-interest theory that says, as we saw earlier, 'that an individual who intends only his own gain, is [. . .] led by an invisible hand to promote [. . .] the public interest' (Hardin 1968, 1244). According to this analysis, climate change is an example of a collective action problem and a social dilemma that highlights the lack of consistency between (1) what is reasonable for each rational agent to do and (2) what is reasonable for the agent within a group to do. Contrary to what self-interest theory predicts, social dilemma theory claims that there is a gap between the individual and the collective level of benefits. Nonetheless, social dilemma theories presume rational self-interest as the driver of choices.

Like self-interest theory, social dilemma theory proposes an explanation of why individual agents fail to protect the earth's climate adequately. This can either be understood in the sense that individual rational agents do face a significant choice in regard to climate change but fail to choose in favour of the climate. Alternatively, it can be interpreted as a theory that does not provide a normative significant choice with regard to climate change because it is methodologically and theoretically precluded that the rational agents can favour actions with high collective and social benefits.

Before proposing which interpretation is the most appropriate, let us look into what a social dilemma is more precisely. Two properties normally define a social dilemma: (1) each individual receives a higher payoff for a socially defecting choice (e.g., having additional children, using all available energy, polluting his or her neighbourhood) than for a socially cooperative choice, no matter what the other individuals in society do, and (2) all individuals are better off if all cooperate than if all defect (Dawes 1980, 169).

Hardin argued that people's freedom to give birth to children runs the risks of lowering the general quality of life of every human being. Public interest is not obtained, according to this reasoning, by aggregation of individuals' personal gains. On the contrary, personal gains deplete the common pool from which the gains are collected. Consequently, the tragedy of people's freedom, as Hardin understood it, captures the moment where 'the inherent logic of the commons remorselessly generates a collapse of the total system in which the individual acts occur' (Hardin 1968, 1244).

Whilst the conclusions of social dilemma theory differ from the conclusions of the self-interest theory, the assumptions about rational self-interest are maintained. The paradox is that it would be in the rational agent's interest to cooperate on climate mitigation and adaptation, for example, but for various reasons the rational agent faces difficulties in doing what would be best for her and the group. The difficulty for rational agents to pursue collective benefits is often conceptualised as a 'prisoner's dilemma'. This challenge illustrates that 'individually rational strategies lead to collectively irrational outcomes' (Ostrom 1990, 5).[5] In prisoner's dilemma scenarios, each person has a dominant strategy to defect cooperation because it is assumed that the person is better off choosing the strategy to defect no matter what the other person chooses (Ostrom 1990, 4–5).

Snatch dilemmas

The social dilemma literature has developed in several ways:[6] over the course of two phases, it has developed into a more relaxed set of assumptions regarding rationality and (short-term) self-interest as the driver of people's choices. The first phase was initiated by Herbert A. Simon with what he called *bounded rationality*, which is the idea that people make choices and decisions under suboptimal conditions. They do not have complete knowledge, and emotions and irrationality may impact their choices and decisions (Simon 1957). The second phase was initiated by Elinor Ostrom, who assumes that not only do people have incomplete knowledge, they also act within a complex network of social norms and institutions that influences their rational incentives, choices, and acts (Ostrom 1990, 2005).[7]

Let us see whether Ostrom's variant of the social dilemma theory provides a more promising approach to rational choices, which allows that rational agents face a normatively significant choice with regard to climate change by being capable of choosing freely to respond in a climate-friendly or unfriendly manner. Within this theoretical frame, people's choices are not only motivated by rational self-interest but also by moral and social norms.

The prisoner's dilemma assumes a set of artificial conditions: without communicating with each other, two people with preferences for self-interest must make a decision that depends on the other person's decision. Recent versions of the social dilemma theory try to produce a theory that is closer to real-world scenarios. One suggestion is the *repeated many-person dilemmas*, which define more complex dilemmas and the rational choice situations than self-interest and classic social dilemma theory do. One dilemma of this kind is called the *snatch dilemma*

Table 4.1 Snatch dilemma I

Agent B	Agent A		
	1)		2)
1)	Exchange: $\pi_A = 15$ $\pi_B = 15$ $\pi_C = 30$		Snatch: $\pi_A = 5$ $\pi_B = 20$ $\pi_C = 25$
2)	Snatch: $\pi_A = 20$ $\pi_B = 5$ $\pi_C = 25$		No exchange: $\pi_A = 10$ $\pi_B = 10$ $\pi_C = 20$

(Ostrom 2005, 36). Let us consider whether this dilemma provides reasons for believing that rational agents face a normatively significant choice with regard to climate change.

In contrast to the prisoner's dilemma, the snatch dilemma accounts for the reaction that agent A is expected to have to the choices and behaviour of agent B. The key aim of the dilemma is to understand why agents are incentivised to snatch. Consider one version of the snatch dilemma in which agent A produces 10 units of chickens and agent B produces 10 units of potatoes. Agents A and B would benefit from exchanging some chickens and some potatoes (see Table 4.1).

The reason for this benefit is that the second commodity of which agents A and B initially had none has a marginal value that is twice as high as the value of the first commodity that they had prior to the exchange (Ostrom 2005, 35–36). Note here that the *exchange* of goods increases not only the individual outcomes but also the collective outcome as well.

The following calculation explains the exchange issue: if one *exchange* of goods occurs, agents A and B obtain values corresponding to 15 units each. If *no exchange* occurs, agents A and B keep 10 units each. Consider the case in which agent A snatches the second commodity from agent B. Then, agent B will have only 5 units, whereas agent A will obtain a value corresponding to 20 units.[8] According to the theoretical assumptions about self-interest, agents will snatch as long as they believe they will gain a positive marginal utility per unit (Hardin 1968, 1244).[9]

This situation occurs 'whenever the private returns to each participant are greater than their share of a joint return no matter what other participants do' (Ostrom 2005, 37). This means that agent A gets a greater return by snatching instead of by exchanging commodities. The snatching dilemma reflects the predictions of *the tragedy of the commons*, where people take more than their fair share of resources. What the snatch dilemma adds to the calculation is the marginal cost/benefits of snatching.

Let us now consider a modified version of the snatch dilemma that suggests an alternative model of choices. One of the main theses of the modified snatch dilemmas is that the rational agent's incentives cannot be considered independently of the institutional and social context (Ostrom 2005, 123). Consider the following modification of the snatch dilemma (see Table 4.2):

Table 4.2 Snatch dilemma II – modified version

Agent B	Agent A					
	1)			2)		
1) *Exchange*:				*Snatch*:		
	$\pi_A = 15$	$\pi_B = 15$	$\pi_C = 30$	$\pi_A = 5$	$\pi_B = 20 - (\pi^s - \delta^s)$	$\pi_C < 25$
2) *Snatch*:				*No exchange*:		
	$\pi_A = 20 - (\pi^s - \delta^s)$	$\pi_B = 5$	$\pi_C < 25$	$\pi_A = 10$	$\pi_B = 10$	$\pi_C = 20$

In this modified version, the payoffs are the same as in the first snatch dilemma.[10] The dissimilarity is that agents A and B hold a social norm against snatching; for example, they may feel guilty about stealing goods from another person. If agent B snatches the commodity from agent A, the utility that agent B gains should be adjusted for the negative utility agent B gets by violating a social norm ($\pi_B = 20 - (\pi^s(\text{snatch}) - \delta^s(\text{aversion to snatching}))$).

Two points should be stressed. First, the agents in the modified snatch dilemma do not have an incentive to snatch because there is a risk of obtaining the same utility or less compared to the exchange of commodities. This risk is caused by the agent's aversion to snatching from her neighbour. The implication of the modified snatch dilemma is that *exchange* gives the highest personal benefit ($\pi_A = 15$) and not *snatching* as in the first version of the snatch dilemma. Note that the modified snatch dilemma does not evade the assumptions of rational choice and self-interest theory. The implication of the snatch dilemma is that rational agents should change their preference order to prefer *exchange* to *snatching* and *snatching* to *no exchange*.

Second, exchange gives the highest collective benefit ($\pi_c = 30$) in the modified snatch dilemma which thus succeeds in solving the discrepancy between self-interest and collective benefits. In other words, the self-interest theory is here neither indirectly nor directly self-defeating, as was the case for the classic version of self-interest theory. The reason for this is that the modified snatch dilemma argues that the highest collective benefit is what gives the highest personal benefit.

Social context and norms

The basic theoretical assumption of the snatch dilemma is that the incentives and choices of rational agents cannot be understood without considering the context within which the act happens and that the rationality of one choice cannot be estimated by reference to self-interest only.[11] In contrast, the claim is that the rationality of one choice can only be understood if the institutional and social contexts are taken into account. There are several competing theories addressing the importance of institutions and norms, for instance, theories of domain-specific institutions, cultural processes, language, locations, and various deontic reasons (Ostrom 2005, 124–127). When one accepts one of these theories, the threshold

(k) for choosing to *cooperate* rather than to *snatch* is a soft external constraint that cannot be reduced to an estimate of individuals' personal self-interests. The reason for this is that the threshold (k) at which rational agents are expected to cooperate depends on institutional and political factors.

Thus, the calculation in which the expected personal benefits of *snatching* are lower than the expected collective benefits of *cooperating* ($k = \pi_c$ (collective benefits) > (π^s(snatch) – δ^s(aversion to snatching)) cannot be estimated without addressing the influence of legal institutions, political regulation, and social norms.[12] The costs of *snatching* depend on the institutional context and social norms held by the community and the agents.[13] Thus, the choice situation the rational agents face with regard to choosing between to cooperate (collective benefits) or to defect (short-term personal benefits) depends on the social institution and norms that influence what rational agents believe to be rational.

One determining factor of the choice situation is what type of social norms dominate the society within which the rational agent is acting. If we take the point of departure from contemporary Western societies, a set of social norms could be (1) 'individuals attribute less value to benefits that they expect to receive in the distant future, and more value to those expected in the immediate future' (Ostrom 1990, 34) and (2) modern individuals do not have social norms that encourage them to consider future people.[14] One justification for this claim is that societies have changed from rural societies to modern urban cities, which suggests that people's social valuing of future individuals has changed.

Sociological research supports that local dependence and long-term ownership over limited territories have shifted in favour of modern industry, higher mobility, and lower solidarity between generations. For example, the time horizon of modern lives is very different from that of indigenous communities, which often is argued to extend far into the future (Stehr 2007; Rosa and Trejo-Mathys 2013). Ostensibly, indigenous and rural communities have a flat discount rate, which implies that what happens to their children's children has no less value than what happens to the current generation. Thus, indigenous and rural communities make a serious effort to protect the common-pool resources whilst using the local natural resources when fishing, hunting, and farming (Ostrom 1990, 35).

This sociological point supports the thesis provided by Dale Jamieson, among others, that the challenge of modern climate politics is the lack of 'green virtues' (Jamieson 2010).[15] Jamieson claims that the idea of green virtues can be accommodated by the self-interest and rational choice theories because it should be in everybody's interest to sustain the earth's sustainable resources and climate.

Nonetheless, it remains undertheorised within the rational-individualism how the exchange of ideas and values happens between the rational agents and the social context. Common for the different accounts of rational agency discussed earlier is that they aim to provide stable patterns of predictable acts and choices that limit the scope of the agent's freedom and, therefore, his or her fitness to be held responsible for bad outcomes such as climate change. Indeed, both Ostrom's suggestion of the compatibility between rational agency and social norms such as guilt and Jamieson's suggestion of the compatibility between rational agency and green virtues aim to improve the rational choice methodology.

However, these improvements do not provide a positive theory of the strength and significance of the social and institutional context, Moreover, nothing is said about how the agent may influence the social and institutional context. Rational-individualism remains an incomplete theory with regard to what constitute social norms, institutional facts, and green virtues. A different methodological theory is needed in order to comprehend how the social norms influence individual agency and how individual agency may scope social norms. As we shall see in the subsequent chapters, other methodologies provide a more comprehensive and positive account of social and institutional spaces.

Note that this cannot be perceived as a critique of rational choice theory. Indeed, one of the main normative implication of rational choice theory is that in many cases individual agents should not be held responsible for the consequences of their choices. The agents in rational choice theory are not pre-ordained to be evil or climate unfriendly, but they are seen as acting according to a set of rational rules (Plant 2009, 91; S. Miller 2010, 59). However, if our goal is to provide a theory of moral responsibility in which the agents assigned responsibility are presumed fit to be held responsible, rational choice theory does not provide what we need. It is unfortunate that contemporary climate ethics and climate politics implicitly or explicitly favour this model for understanding the moral dilemmas related to climate change, as we shall see in Chapters 5 and 6.

If the arguments offered thus far are on the right track, it is difficult to assign responsibility at the level of individual rational agency. Having considered three different versions of rational agency theory, I conclude that none of them provides a level of agency that is fit to be held responsible for climate change because the agent in many cases does not face a free choice but is bound to what is rationally expected of agents to do. Thus, we can conclude that individual rational agents do not face a normatively significant choice with regard to the problems of *the tragedy of the commons* related to climate change. Indeed, the *tragedy of the commons* relies on a theory in which no agents are relevant moral agents in regard to climate change.

Miscalculation of social benefits

In the previous section, I argued that rational-individualism does not offer a theoretical framework that acknowledges that individual rational agents face a morally significant choice with regard to climate change. I also argued that rational agents do not face a significant choice because their rational choices are dependent on their rationality or the social norms available in the modern societies. In this case, the first condition of fitness for moral responsibility is not satisfied. Although rational agents do not face a significant choice, I continue the investigation of why the rational agents are expected to fail systematically to opt for a climate-friendly solution with high collective benefits.

What are the rational mechanisms that drive the social dilemma? In the rest of this discussion, I investigate why rational agents in the rational choice frameworks are systematically expected to fail to choose in favour of climate friendliness. This is a discussion of the second fitness-condition of responsibility, which concerns whether agents have access to sufficient and relevant knowledge. I also discuss whether the choices of rational agents are dependent on social and insti-tutional contexts. This is a discussion of the third fitness-condition about who has control over the choice situations. Let us begin with the second condition, which claims that in order to hold agents responsible, they should have access to the evidence required for to make normative judgements about the relevant choices (List and Pettit 2011, 158). I discuss whether the methodological framework of rational-individualism succeeds in this regard.

What type of knowledge is presumed to be available to individual rational agents? One way to put the question of knowledge in the rational choice models is to consider the calculation of costs and benefits as a practice of knowledge. In this regard, the social dilemma models discussed earlier reflect a social contract theory in which people consent to a contract if they estimate the personal net benefits to be high. In the Hobbesian contract theory, people subject themselves to a Leviathan in order to gain personal security (Hobbes 1991, Chapter 13). The benefits are secured because the Leviathan will punish those who defect. In the modified snatch dilemma, what stopped B from snatching was not a Leviathan but B's aversion to guilt. The notion of guilt is nonetheless supposed to do the same job as the Leviathan. This means that guilt has a regulating order that stems from people's dislike of violating social norms.

The logic of the modified snatch dilemma is meant to cover all social dilem-mas of public goods and common-pool resources. Social feedback mechanisms such as guilt presume that people know that they have done something wrong, as in the case of snatching from the neighbour. If this is plausible, individual ratio-nal agents have sufficient knowledge with which to estimate benefits and costs. However, the Hobbesian model and the guilt model do not completely resemble the challenges related to intergenerational climate change because the costs of climate change are differently composed compared to those of security and of sus-tainable natural resources upon which Ostrom focuses (Ostrom 1990). The reason for this is that the benefits of cooperating about climate policies will not occur immediately. In contrast to the case of social security, which people benefit from immediately by enjoying, for example, peace, the benefits of avoiding climate change may seem unnoticeable for a long time, say, for 50–100 years.

If we take this time lag into account, the modified snatch dilemma cannot pre-dict that agents will cooperate to mitigate climate change. There may be no incen-tives to contribute to the sustainability of the planet's climate and atmosphere if the costs of climate change (δ_i) are very low in the first 50–100 years, which implies that the personal benefits (π_i) in the same period will be very high. More-over, if people do not know the long-term negative consequences of, for instance, driving a CO_2-emitting car, they will not feel guilty.[16] Hence, by introducing a time gap between benefits and costs, the calculation of both personal benefits adjusted for personal costs ($\pi_i - \delta_i$) can only be arbitrarily estimated at T1. More-over, with regard to climate change, the threshold at which collective benefits are greater than personal benefits minus costs ($\pi_c > (\pi_i - \delta_i)$) is unclear.

The consequence of the time gap is a *systematic miscalculation of collective benefits* at T1, which is explained by the fact that personal benefits occur now (at

T1), whilst collective benefits and costs first occur in the future (at T2). This claim about a systematic miscalculation of collective benefits is justified by the fact that climate sensitivity, global warming, transformation of physical geography, and CO_2 emissions all involve time lags (N. Stern 2010, 40). Add to this that each benefit involves considerable uncertainty. It will be clear that at T2 the estimate of collective benefits relative to personal benefits is both individually self-defeating (because the sum of personal benefits over time is less than the personal costs over time ($\Sigma(\pi_i) < \Sigma(\delta_i)$)) and collectively self-defeating (because collective benefits are thus less than the collective costs ($\pi_c < \delta_c$)).

To summarise, social dilemmas with a significant time gap can explain why it is rational for individuals not to contribute to the collective benefits of the planet's climate and atmosphere. The theoretical explanation is that people systematically miscalculate the level of collective benefits to be less than personal benefits minus costs (i.e., they mistake $\pi_c > (\pi_i - \delta_i)$ for $\pi_c < (\pi_i - \delta_i)$).[17] In other words, in rational choice theory, it is rational for people to act in a way that expects collective benefits to be less than personal benefits. The presumption is, as Arrow argued, that individuals do not have the appropriate measurements to judge future risks and costs, and, therefore, are incapable of making proper probability judgments about future risks (Arrow 1983, 21).

A further challenge is related to choices under uncertainty. 'In many of the most difficult situations, those with high risks but very low probabilities [. . .], the evidence on the relevant uncertainty is extremely small' (Arrow 1983, 21). In deciding what is rational to do, it is impossible to take actual consequences into consideration: one can assess only the likely consequences (Prior 1956, 99). The problem is convincingly summed up by Dietz, Ostrom, and Stern: 'Devising ways to sustain the earth's ability to support diverse life, including a reasonable quality of life for humans, involves making tough decisions under uncertainty, complexity, and substantial biophysical constraints as well as conflicting human values and interests' (Dietz et al. 2003, 1907).[18]

In order to introduce some clarity to the debate on the systematic miscalculation of collective benefits, three properties should be distinguished: (1) uncertainty, (2) unpredictability, and (3) lack of knowledge (Ostrom 1990, 33). To explicate the three properties, let us consider the following *agent A* example. (1) Sixty years ago, agent A built a plant constructed with a pipeline that released a polluted steam from chemical production into the air. The local government approved the construction plan of the plant. Every year, 100 tonnes of polluted steam are emitted by the plant. (2) The agent A knew that the steam was polluted and dangerous for the plant's employees (*certainty*). Hence, there was a predictable risk that the emissions of polluted steam could be linked causally 40 years later to the contamination of the clean water supplying the employees, their families, and the local communities (*predictability*). (3) However, agent A did not know that the emissions of polluted steam on an aggregated level 100 years later would be linked causally to substantial global climate change (*lack of knowledge*).[19]

This example illustrates that collective benefits that are subjected to a time gap make it notoriously difficult to estimate their value in rational choice theory.

Therefore, the modified snatch dilemma cannot accommodate social cooperation on climate-related issues. The snatch dilemma allows for the possibility of calculating personal benefits compared to public benefits but not in the case of climate change, where there is a time gap between costs and benefits.

By so arguing, the theory of individual responsibility of rational agents does not allow for access to the relevant knowledge with regard to climate change because of the systematic miscalculation of collective benefits. Because rational agents are expected to miscalculate costs and benefits, they thwart what might be in their own interests. So interpreted, the second fitness-condition of responsibility addressing the access to relevant knowledge is not satisfied. Within the rational set of assumptions, rational agents can be held responsible for something we would expect them to do. If one thinks this is unreasonable but at the same time believes it is unreasonable to let individuals off the hook, another solution might be to relax the theoretical assumptions of self-interest and rational choice theory.

Based on this analysis, I conclude that the theory of individual responsibility does not successfully argue that the presumed concept of rationality and level of agency makes rational agents fit to be held responsible for climate change. The theory fails to satisfy the first two criteria for being fit for responsibility: (1) that individual agents have the necessary cognitive and judgemental capacity to understand and assess evidence of climate change and, therefore, also (2) that individual agents are faced with a morally significant set of choices. In the next section, I discuss whether the theory of individual responsibility fails the third condition as well – namely, to prove that individual agents have the control required for choosing between the options.

Social behavioural spaces

Having argued that within the framework of rational choice individual agents are systematically expected to miscalculate the collective benefit of environmental sustainability, let us now discuss to what degree rational agents can be thought to have control over choice situations related to the systematic miscalculation of collective benefits. Who has control over the acts depleting resources if (1) the depletion is uncertain and unpredictable and if (2) individual rational incentives within the present social behavioural space conflict with what is individually and collectively rational in the long run? Choices in this context are not merely a question of rational incentives but also about how the social and institutional landscape is structured.

In Ostrom's understanding of rational choice theory, institutions shape what is rational for agents to do. According to Ostrom, incentive structures vary according to social norms, political regulations, and other institutional arrangements.[20] These behavioural patterns, which are culturally, legally, politically, or economically framed, are thought to influence individual incentives and what is perceived as rational to do.[21] As argued earlier, rational agent theory does not provide a positive account of how societal patterns are organised. For the sake of argument, let us accept the minimal assumption that a layer of societal norms or structures shapes the behavioural space within which the rational agents act. For present purposes, we can thus remain neutral as to whether it is culture, institutions, or other social entities that form the scope of behavioural patterns.

A behavioural pattern consists of all the movements and acts an individual agent could possibly make at T1. For every time T, the behavioural pattern limits the agent's options and alters 'the set of alternatives facing the agent, not the agent's attitude towards the (original) alternatives' (Taylor 1982, 20–21). This means that individuals are not completely in control of what is reasonable and rational for them to do within a given behavioural pattern.

The theory of behavioural spaces presumes that the agent is rational and intends to make rational choices, but that the scope of possible candidates for choices is limited by the surrounding behavioural space. Moreover, the restricted behavioural space affects the estimated probability of the outcomes of different actions. As Brian Barry suggests,

> To choose between options sensibly, you need to understand the nature of the options. There will normally be a range of possible outcomes arising from each choice, and you need to have at least some estimate of their relative likelihoods.
>
> (Barry 2005, 137)

Suppose now that the different possible movements and acts are not equally distributed within a behavioural pattern at T1. Of all possible acts, let us say that 90 percent will have the same effect; only 10 percent will have a different outcome (Bennett 1995, 93–95). Whilst this may seem to be an unreasonable theoretical assumption, many real-world examples are constructed in this way. Consider this example: 'A' may want to buy an environmentally friendly car. Let us say that there are ten different cars on the market that that A can afford. Nine out of ten cars are equally environmentally unfriendly, whereas the tenth car has a more environmentally friendly impact. Only by buying that car out of the ten will have a different outcome. Within this behavioural pattern, it is more likely that A will choose one type of behaviour from the 90 percent area rather than from the 10 percent area. It is most likely that A will buy an environmentally unfriendly car when nine out of the ten cars have the same negative environmental impact. One explanation for this behaviour is that it seems rational for agents to do what is most likely given the present social and political arrangements.

The challenge can be formulated in a different way. The question is what we can expect rational agents to do if the present behavioural pattern makes environmentally unfriendly consumption more likely. Is it fair to expect ordinary and rational agents to do something that is very unlikely or difficult to fulfil within a present behavioural pattern? If the availability of environmentally friendly cars is very low, the technology is poorly developed, and the cars are extremely costly, we cannot expect agents to give up their environmentally unfriendly cars for which availability is very high and technology is extremely developed, and which are inexpensive compared to the environmentally friendly cars.[22] Arguably, rational choices depend on (1) insights into the possibilities within the behavioural space and on (2) reasonable estimates of the choice' relative outcomes. So conceived, it is not the rational agent that controls the choice situation, which provides evidence that the rational agents are unfit to be held responsible according to the third fitness criterion.

Who are the polluters?

Until now I have discussed the concept of rationality of individual responsibility for climate change. It has been presumed that the relevant polluters are rational individuals. In this section, I argue that despite the fact that rational agency is not a suitable framework for making agents fit to be held responsible, the theory of individual responsibility when understood as the PPP cannot be disregarded without further justification. The earlier critique of rational agency theory is not an argument against the validity of the PPP. Although the principle is frequently linked to rational agency, I argue that the PPP sustains its plausibility despite the fact that the individual level of rational agency is unfit for responsibility.

The PPP exists in different versions in the scholarly debate. One version establishes that past emitters should economically compensate the victims exposed to emissions (Page 2006, 53; Jagers and Duus-Otterström 2008, 578). Another version establishes that

> [w]hen a party has in the past taken an unfair advantage of others by imposing costs upon them without their consent, those who have been unilaterally put at a disadvantage are entitled to demand that in the future the offending party shoulder burdens that are unequal at least to the extent of the unfair advantage previously taken, in order to restore equality.
>
> (Shue 1999, 534)

Generally, proponents of the PPP agree on the need to draw a causal and moral link between the environmentally harmful acts of the polluters and the consequences of affecting someone negatively. We could call this a *perpetrator-victim-model*. Most people would agree that it is reasonable to claim that a murderer should be held accountable for the killing of a victim. Likewise, most people would agree that it is reasonable that an oil company that spills oil on farmers' crops should compensate the farmers for the losses they may suffer.

Despite the plausibility of this, many argue that the PPP fails in cases of intergenerational climate change. Two identity problems challenge the PPP. The first challenge, which is commonly recognised, is that the original polluters may be dead at the moment of compensation and therefore unable to restore the unfair and unequal advantages that they have received.[23] The second identity challenge concerns a non-identity problem of the victims. The PPP assumes that it is feasible to identify the polluters that should compensate the victims of high emission rates. Likewise, the PPP presumes that the victims are identifiable. There is, however, a non-identity problem on the side of the climate change victim as well. Many victims may be dead at the moment of compensation. Others may have adapted to the changes, for instance, by building a new house. It would be unfair not to count them as victims just because they can afford a new house. Other obstacles to a fair assessment of responsibility may be impossible to identify. Hence, it is unclear to whom polluters owe compensation.[24]

Note that the two non-identity problems do not demonstrate that the PPP applied at the level of individual polluters is self-defeating or inconsistent. The two problems of non-identity reveal only that the PPP is a politically unfruitful strategy because the victims will not be economically compensated by the original polluters (as the PPP prescribes). Thus, the PPP fails if the aim is to match a singular identifiable individual perpetrator with a particular number of identifiable victims. However, the polluter to which the PPP refers may not be an individual: it can be a group agent, private company, a state, or an international cooperation (Caney 2010, 26–27). If we allow for collective polluters such as companies, states, and international organisations, the challenges of the two non-identity problems in most cases will dissolve. Most private companies and states maintain an identity across time, which makes the non-identity problems less severe.[25]

For the last decade, it has become common to refer to the PPP as a model for individual responsibility in the political and philosophical literature on climate change (Gardiner 2004; Parfit 2010; Broome 2012). The philosophical debate on the PPP tend to take the polluters to be private individual agents (e.g., Gardiner 2004; Broome 2012) or states (e.g., Shue 1999). However, it should be noted that the economic theory of the PPP originally defined the polluters to be industrial plants and private corporations. The PPP was initially formulated as a principle in the economic literature addressing the issue of internalising negative externalities (i.e., environmental harm and climate change) in the prices of products sold on the market.

One of the first international organisations to promote the concept of the PPP was the Organisation for Economic Co-operation and Development (OECD). In 1972, the PPP was incorporated for the first time in the OECD's *Council Recommendation on Guiding Principles concerning International Economic Aspects of Environmental Policies* (Nash 2000; OECD 2002, 6). The initial principle said that 'the polluter should bear the costs of pollution prevention and control measures', and that 'measures decided by public authorities should ensure that the environment is in an acceptable state' (OECD 1992, § 1.1). Later, the PPP came to include several economic instruments (OECD 2002, 6) and full compensation costs (OECD 1992, § 1.1).[26] The concept of the PPP has been incorporated in a number of other international environmental instruments and texts. Many OECD member countries and the European Commission have accepted the concept of the PPP underlying their environmental policies (OECD 2002, 10). Furthermore, the World Bank and UNFCCC recognise the principle.

The economic justification for the PPP is the standard theory of negative externalities. This theory holds that GHG emissions damage others 'at no cost to the agent responsible for the emissions' (N. Stern 2006, 4). In order to obtain efficient economic competition and the efficient use of resources, the costs of the externality caused by the polluters should be internalised – i.e., paid by the polluting producers or consumers. Proponents of internalising negative climate externalities claim that the environmental costs caused by a product's production chain should be reflected in the product's price, which requires political regulation and monitoring. There has been a recent wave of environmental legislation and political instruments in most OECD countries (Sagoff 2008, 9) that ensure, among other things, the internalisation of environmental externalities. Pollution charges and taxes, for example, aim to reduce the use of nitrates and pesticides in agriculture (OECD 2002, 23). The overall picture is that these instruments have been quite successful: for example, air pollution in the US has fallen to the lowest level ever recorded (Sagoff 2008, 5).

Internalisation of environmental costs can be addressed for polluters such as industrial plants and private corporations. Products that have an environmentally harmful production or decomposition cycle should be more expensive in order to incentivise more environmentally friendly production, materials, and resources. However, if polluters are defined as individual agents, the theory of internalisation of negative externalities crumbles: only burden sharing and victim compensation are left as possible actions against polluters. The two latter points are arguments in favour of the idea that the PPP should not be extended to cover the responsibility that individuals bear and that the rejection of the PPP at the level of individual polluters and victims does not allow for disregarding the PPP at the level of collective group agency.

Environmental harm and climate change

One further point should be noted with regard to the critique of the PPP. Challenging the relevance of the PPP in the cases of intergenerational climate change should not lead us to conclude that the PPP is irrelevant in cases of environmental pollution. The PPP succeeds when a perpetrator and victim are matched in such cases. Hence, it is reasonable to suggest distinguishing local *environmental pollution* from spatially and temporally diffused *climate change*.[27] This may be a common-sense distinction but is nonetheless frequently neglected in the literature on climate ethics. Local environmental pollution consists of perceptible harms that are contemporary and local.[28]

In contrast, spatially and temporally diffused climate changes are more loosely connected to the contemporary and local context and are caused by countless and imperceptible harms over a long period of time. Here, because the cause and effects are diffused and not closely interlinked, perpetrators and victims are not easily identifiable. To clarify the distinction, let us consider the agent A example again. A has contaminated the local water supply. The perpetrator is clearly A, and the victims are clearly the group of people drinking the contaminated water. In the case of climate change, it is not possible to track A's contribution to consequences in 100 years – e.g., rising sea levels. Moreover, the victims of climate change to which A owes compensation are impossible to identity.

If we take the *two identity problems* into account and at the same time accept the distinction between local environmental harm and global climate change, we can determine the conditions under which the PPP fails and succeeds. The PPP succeeds (1) in cases of environmental pollution caused by individual agents and (2) in cases of environmental pollution caused by private companies and states. Furthermore, the PPP succeeds (3) in cases caused by private corporations. The PPP fails, however, (4) in cases of spatially and temporally diffused climate caused by individual agents. If (1), (2), and (3) are conflated, the challenges of climate and environmental pollution are misconstrued.

So conceived, the PPP is compatible with the assignment of responsibility to single individuals, private corporations, or other group agents. Therefore, a conclusive suggestion is to disentangle the economic and rational-theoretical presumptions on the one hand, and the theory of the PPP, which assigns moral responsibility to the agents that contribute to environmental harm and climate change, on the other. This manoeuvre requires a level of political governance that is powerful enough to regulate the economic incentives in a green direction. A comprehensive account of climate governance is more fruitfully addressed in the non-reductionist institutionalism (established in Chapter 3). This methodology allows for a more complex theory of agency, rationality, and, therefore, responsibility.

Conclusion

I have focused in this chapter on the fitness-conditions that are fundamental to understand the ought-assignment to particular agents, in this case rational individual agents. The three fitness-conditions should be satisfied by virtue of the specific assignment of moral responsibility for climate change based on the fact-sensitive account of moral responsibility elaborated on in Part 1. In this chapter, I have discussed three ways to approach the dominant view that the problem of climate change is the rational agents that cause *the tragedy of the commons* to happen. I have tested these approaches by applying the fitness-conditions of moral responsibility – namely, whether the agents face a normatively significant choice, possess sufficient knowledge, and have control over the choice. These conditions have been applied to climate change.

I have argued that the theory of self-interest and social dilemma theory presumes a concept of rational agency that does not allow the agents to face the morally significant choice of climate friendliness. Moreover, rational agents are unfit to be held responsible because their judgemental capacity is insufficient to calculate the benefits of environmental sustainability. And, finally, in most cases, rational agents do not have control over the societal infrastructure, which makes environmentally friendly choices hard to expect. The chapter concludes, therefore, that the methodology of rational-individualism provides a framework in which rational agents are unfit to be held responsible for climate change. The framework operates with no relevant level of moral agency for climate action.

Finally, I have shown that this does not imply that the PPP fails or that rational agents are unfit to be held responsible for the environmental harm they may cause.

Notes

1 There are different versions of the self-interest theory and different concepts of how people's self-interests are constructed and what is in people's self-interest. Here, I set these dissimilarities aside.

2 A theory is indirectly self-defeating 'when it is true that, if someone tries to achieve his theory-given aims, these aims will be, on the whole, worse achieved' (Parfit 1984, 5). The theory is not directly self-defeating if it remains rational for one to aim to achieve the theory-given aims (Parfit 1984, 13). Furthermore, self-interest theories are directly collectively self-defeating 'when it is true that, if all of us successfully follow the self-interest theory, we will thereby cause our theory-given aims to be worse achieved than they would have been if none of us had successfully followed the self-interest-theory' (Parfit 1984, 53).

3 In Chapter 5, I return to the question of value theory. In this section, I discuss the relevance of economic discount rates compared to social norms.

4 In the economic literature, the collective benefits ignorance is called negative externalities and market failures, see, e.g., Buchanan (1962). Stern argued that 'climate change is the greatest market failure the world has ever seen' (N. Stern 2010, 40).

5 The prisoner's dilemma explains the occurrence of social co-ordination problems when (1) agents have complete information, e.g., they know the full structure of the game and the plausible outcomes and (2) communication between the agents is forbidden or impossible. Consider the example in Matrix X, which follows. Two people are questioned separately about a joint crime. The attorney suggests two alternatives to each person: to confess to the crime or not to confess. If neither confesses, they will both receive two years' punishment for some minor illegality [B, B].

Matrix X	Agent II		
Agent I		A (Confess)	B (Not-confess)
	A (Confess)	10 , 10	12 , 0
	B (Not-confess)	0 , 12	2 , 2

If they both confess, they will receive less than the most severe sentence [A, A] (see Parfit 1984, 56ff.; Ostrom 1990, 3–28, 217; Kurrild-Klitgaard 2010, 342). For both agents, the initial order of preference is [B,A], [B,B], [A,A], and [A,B]. The prisoner's dilemma game illustrates the problem that it is impossible for an agent to choose the decision with the best outcome if unable to take into account the decisions of the other agents. Agent I's dominant strategy is to play 'not-confess', but the outcome of the game still depends on the choice of agent II. Given that neither agent I nor II knows what the other agent plays, the initial preference ordering is unreliable and may be self-defeating.

6 One example is *rational choice institutionalism* (RCI). RCI focuses on how individual agents can use institutions strategically (Hall and Taylor 1996). The institutionalism I am interested in here is the one that establishes the social and institutional environment and context of individual activities. Hence, RCI is for present purposes set aside.

7 Although Simon (1957) and Ostrom (1990) remain within the framework of methodological rationalistic individualism, the broader the package of preferences and norms makes it unlikely that the models for agency can be constructed by an individualist and rational methodology alone. By addressing the social and collective aspects of people's interest and motives, Simon and Ostrom pave the way for expanding the explanatory level to include collective, social, and institutional kinds of agency. For the sake of argument, let us for now accept the individualistic interpretation of how to perceive

social norms. In the following section, I discuss Ostrom's critique and development of the rational choice and self-interest theory and their account of an agent's reasoning and choices.

8 The value of 20 units may be lower due to a decrease in marginal utility of the last 5–10 units.

9 An exchange situation in which the highest level of collective utility is not reached is called a 'Pareto-inferior exchange'.

10 Following Ostrom, the preference function for agent B can be represented by: $u_2 = \pi_2 - \delta^s$, where π_2 is payoff obtained; δ^s is the decrease in the value of π_2 for violating the social norm (Ostrom 2005, 122).

11 Note that the snatch dilemma is not a test of the prisoner's dilemma; it emphasises instead the limited use of the prisoner's dilemma in social dilemmas with common-pool resources (Ostrom 1990, 183–184).

12 Questions of distribution are here set aside.

13 By applying this institutional analysis to several empirical studies, Ostrom demonstrates that governing the commons sometimes succeeds and sometimes fails (Ostrom 1990, 29). The explanation for varying success in governing can be found in the particular institutional arrangements.

14 See Stehr (2007). Another way of understanding intergenerational solidarity argues that the benefit of future generations increases if climate change mitigation policies are postponed until more efficient technologies are provided.

15 Moreover, there is a vast literature on modern human being's alienation from Mother Nature and Gaia (Lovelock 2009).

16 Note that it is uncertain to what extent guilt and shame are effective in motivating people to engage in environmentally friendly behaviour (see Swim et al. 2009 for further discussion).

17 To avoid the discussion about to what extent it is fair to give priority to human entities over non-human entities, $\pi_{(nature)}$ and $\pi_{(animals)}$ are not taken into consideration. Possible positive effects of climate change are also set aside. Some evolutionary biologists argue that the changes of the planet's climate and atmosphere have positive consequences for new species and better living conditions for certain types of animals (Jablonski 2001, 5359; de Perthuis 2011, 36).

18 Note that whilst current predictability of climate change is high, the same cannot be claimed for pre-1990 emissions (Caney 2005, 761–762; Singer 2010).

19 Whereas the first two outcomes are examples of environmental pollution, the latter is an example of spatially and temporally diffused climate change, which I distinguish at the end of this chapter.

20 Self-interest and rational choice theory is frequently connected to a market theory, in which it is claimed that people's estimation and deliberation errors can be defeated by the market. I consider this suggestion in greater detail in Chapter 5. Moreover, many believe that people have at least two sets of motives: one related to economics and self-interest and one related to moral questions. If we accept this, one important challenge is to establish institutional arrangements that appeal to both sets of motives (Frey and Jegen 2000).

21 See, e.g., Taylor (1982); Bennett (1995, 92ff.), Cohen (1997), Ostrom (1990), Ostrom (2005), and Geuss (2008, 135–150).

22 To what extent it is impossible or too costly for someone to buy an environmentally friendly car depends, of course, on her income. Environmentally, friendly cars are less costly for wealthy people compared to poor people. This issue of income inequality is here set aside.

23 Dale Jamieson (2010, 83) also makes this point.

24 The non-identity problem hinges more generally on the risk that 'after a few generations, and depending on which option we choose, completely different sets of people

come into existence and these sets of people will in a sense owe their existence to this prior choice of option (they would not have been born if that particular option had not been chosen)' (Page 2006, 134; see also Parfit 2010).

25 There may be some special cases of collective non-identity problems, but I set these cases aside.

26 Another aspect of the PPP is the concept of extended producer responsibility (EPR), which places responsibility for a product's end-of-life impacts on the original producer (OECD 1996, 4). 'EPR is employed by governments as a strategy to transfer the costs of municipal waste management from local authorities to those actors (i.e., the producers) most able to influence the characteristics of products which can become problematic at the post-consumer stage: waste volume, toxicity, and recyclability' (OECD 1996, 6).

27 To be more specific, climate change consists of the long-term aggregation of environmental pollution. The difference between the two presumes that it makes sense to differentiate between short-term and long-term effects. I am aware that this distinction is idealised, but as an analytical tool, it suffices for present purposes.

28 One example could be the British Petroleum oil spill in the Gulf of Mexico in 2010. In this case, because cause and effects are closely interlinked, the perpetrators and victims are easily identifiable.

References

Arrow, K. J., 1983. Behavior under uncertainty and its implications for policy. *In:* B. P. Stigum and F. Wenstøp, eds. *Foundations of utility and risk theory with applications.* Dordrecht: D. Reidel, 19–34.

Barry, B., 2005. *Why social justice matters.* Cambridge: Polity Press.

Bennett, J., 1995. *The act itself.* Oxford: Oxford University Press.

Broome, J., 2012. *Climate matters: ethics in a warming world.* London: W.W. Norton.

Buchanan, J., 1962. Externality. *Economica*, 29 (116), 371–384.

Caney, S., 2005. Cosmopolitan justice, responsibility, and global climate change. *Leiden Journal of International Law*, 18 (4), 747–775.

Caney, S., 2010. Climate change, human rights and moral thresholds. *In:* S. Humphreys, ed. *Human rights and climate change.* New York, NY: Cambridge University Press, 69–90. Reprinted in 2010. *In:* S. M. Gardiner, S. Caney, D. Jamieson, and H. Shue, eds. *Climate ethics: essential readings.* Oxford: Oxford University Press, 163–180.

Cohen, G. A., 1997. Where the action is: on the site of distributive justice. *Philosophy and Public Affairs*, 26 (1) (Winter), 3–20.

Cox, G. W., 2004. Lies, damned lies, and rational choice analyses. *In:* I. Shapiro, R. M. Smith, T. E. Masoud, eds. *Problems and methods in the study of politics.* Cambridge: Cambridge University Press, 167–185.

Dawes, R. M., 1980. Social dilemmas. *Annual Review of Psychology*, 31 (91), 169–193.

De Perthuis, C., 2011. *Economic choices in a warming world.* Cambridge: Cambridge University Press.

Dietz, T., Ostrom, E., and Stern, P. C., 2003. The struggle to govern the commons. *Science*, 302 (5652), 1907–1912.

Elster, J., 1982. Sour grapes: utilitarianism and the genesis of wants. *In:* A. Sen and B. Williams, eds. *Utilitarianism and beyond.* Cambridge: Cambridge University Press, 219–238.

Elster, J., 2009. *Reason and rationality.* Princeton, NJ: Princeton University Press.

Frey, B. S. and Jegen, R., 2000. *Motivation crowding theory: a survey of empirical evidence.* Zurich IEER Working Paper No. 26; CESifo Working Paper Series No. 245. Zurich: University of Zurich.

Gardiner, S. M., 2004. Ethics and global climate change. *Ethics*, 114 (3), 555–600.

Geuss, R., 2008. *Philosophy and real politics*. Princeton, NJ: Princeton University Press).

Hall, P. A. and Taylor, R., 1996. Political science and the three institutionalisms. *Political Studies*, 44, 936–957.

Hardin, G., 1968. The tragedy of the commons. *Science*, 162 (3859), 1243–1248.

Hobbes, T., 1991/1651. *Leviathan*. R. Tuck, ed. Cambridge: Cambridge University Press.

Hurley, S., 2011. The public ecology of responsibility. *In:* C. Knight and Z. Stemplowska, eds. *Responsibility and distributive justice*. Oxford: Oxford University Press, 187–215.

Jablonski, D., 2001. Lessons from the past: evolutionary impacts of mass extinctions. *PNAS*, 98 (10), 5393–5398.

Jagers, S. C. and Duus-Otterström, G., 2008. Dual climate change responsibility: on moral divergences between mitigation and adaptation. *Environmental Politics*, 17 (4), 576–591.

Jamieson, D., 2010. Climate change, responsibility, and justice. *Science and Engineering Ethics*, 16 (3), 431–445.

Kurrild-Klitgaard, P., 2010. Exit, collective action and polycentric political systems. *Public Choice*, 143 (3–4), 339–352.

List, C. and Pettit, P., 2011. *Group agency: the possibility, design, and status of corporate agents*. Oxford: Oxford University Press.

Lovelock, J., 2009. *The vanishing face of Gaia: a final warning*. New York, NY: Basic Books.

Miller, S., 2010. *The moral foundations of social institutions: a philosophical study*. Cambridge: Cambridge University Press.

Miller, S., 2011. Collective responsibility, epistemic action and climate change. *In:* N. Vincent, I. van de Poel, and J. van den Hoven, eds. *Moral responsibility: beyond free will and determinism*. Dordrecht: Springer, 219–246.

Nash, J., 2000. Too much market? Conflict between tradable pollution allowances and the 'polluter pays' principle. *Harvard Environmental Law Review*, 24 (2), 465–535.

OECD (Organization for Economic Co-operation and Development), 1992. *General distribution OCDE/GD (92) 81. The polluter pays principle: OECD analyses and recommendations*. Paris: OECD.

OECD (Organisation for Economic Co-operation and Development), 1996. *Pollution prevention and control extended producer responsibility in the OECD area, Phase 1 report. Legal and administrative approaches in member countries and policy options for EPR programmes, OECD/GD(96)*. Paris: OECD.

OECD (Organisation for Economic Co-operation and Development), 2002. *Joint Working Party on Trade and Environment, COM/ENV/TV(2001)44/FINAL. The polluter pays principle as it relates to international trade*. Paris: OECD.

Ostrom, E., 1990. *Governing the commons: the evolution of institutions for collective action*. Cambridge: Cambridge University Press.

Ostrom, E., 2005. *Understanding institutional diversity*. Princeton, NJ: Princeton University Press.

Page, E., 2006. *Climate change, justice and future gener*ations. Cheltenham: Edward Elgar.

Parfit, D., 1984. *Reasons and persons*. Oxford: Clarendon Press.

Parfit, D., 2010. Energy policy and the further future: the identity problem. *In:* S. M. Gardiner, S. Caney, D. Jamieson and H. Shue, eds. *Climate ethics: essential Readings*. Oxford: Oxford University Press, 112–121.

Parfit, D., 2011. *On what matters*. Oxford: Oxford University Press.

Plant, R., 2009. *The neoliberal state*. Oxford: Oxford University Press.

Prior, A. N., 1956. The consequences of actions. *Proceedings of the Aristotelian Society, Supplementary Volumes*, 30, 91–119.

Rawls, J., 1999. *The law of peoples: with the idea of public reason revisited*. Cambridge, MA: Harvard University Press.

Rosa, H. and Trejo-Mathys, J., 2013. *Social acceleration: a new theory of modernity*. New York, NY: Columbia University Press.

Sagoff, M., 2008. *The economy of the earth: philosophy, law and the environment*. New York, NY: Cambridge University Press.

Shue, H., (1999). Global environmental and international inequality. *International Affairs*, 75 (3), 531–545. Reprinted in 2010: *In:* S. M. Gardiner, S. Caney, D. Jamieson, and H. Shue, eds. *Climate ethics: essential readings*. Oxford: Oxford University Press, 101–111.

Sidgwick, H., 1981/1907. *The methods of ethics*. 7th edn. Indianapolis, IN: Hackett.

Simon, H., 1957. *Models of man: social and rational: mathematical essays on rational human behavior in a social setting*. New York, NY: Wiley.

Singer, P., 2010. One atmosphere. *In:* S. M. Gardiner, S. Caney, D. Jamieson, and H. Shue, eds. *Climate ethics: essential readings*. Oxford: Oxford University Press, 181–199.

Sinnott-Armstrong, W., 2010. It's not my fault: global warming and individual moral obligations. *In:* W. Sinnott-Armstrong and R. Howarth, eds. *Perspectives on Climate Change, Vol. 5*. Bingley, UK: Emerald Group Publishing, 221–253.

Smith, A., 1790/2009. *The theory of moral sentiments*. 6th edn. New York, NY: Penguin.

Stern, N., 2006. What is the economics of climate change? *World Economics*, 7 (2), 1–10.

Stehr, N., 2007. [The morality of the market. A societal theory.] *Die Moralisierung der Märkte: Eine Gesellschaftstheorie*. Frankfurt am Main: Suhrkamp.

Stern, N., 2010. The economics of climate change. *In:* S. M. Gardiner, S. Caney, D. Jamieson, and H. Shue, eds. *Climate ethics: essential readings*. Oxford: Oxford University Press, 39–76.

Swim, J., Clayton, S., Doherty, T., Gifford, R., Howard, G., Reser, J., Stern, P., and Weber, E., 2009. *Psychology and global climate change: addressing a multi-faceted phenomenon and set of challenges*. A Report by the American Psychological Association's Task Force on the Interface Between Psychology and Global Climate Change. American Psychological Society. Available from: www.apa.org/science/about/publications/climate-change-booklet.pdf [Accessed 12 April 2017].

Taylor, M., 1982. *Community, anarchy and liberty*. Cambridge: Cambridge University Press.

Vanderheiden, S., 2011. Climate change and collective responsibility. *In:* N. Vincent, I. van de Poel, and J. van den Hoven, eds. *Moral responsibility: beyond free will and determinism*. Dordrecht: Springer, 201–218.

Williams, B., 1973. A critique of utilitarianism. *In:* J. J. C. Smart and B. Williams, eds. *Utilitarianism for and against*. New York, NY: Cambridge University Press, 77–150.

5   Fitness-conditions of group agents

In the previous chapter, we discussed various economic-rationalist aspects of agency related to the climate change that are understood as *the tragedy of the commons*. Another important aspect of climate change is *the tragedy of the few*, which addresses the set of problems related to how to govern environmental harm and climate change politically.

In the next two chapters, I apply the non-reductionist methodological approach to institutionalised group agency and individual agency elaborated on in Chapter 3. This approach relaxes the economic rationality of agency and points to other equally important public reason aspects of agency which may be nurtured, for example, by education, free media, and democratic participation and representation. The central and most important purpose of the following two chapters is to examine whether the non-reductionist methodological approach to agency allows for a framework in which institutionalised groups are fit to be held morally responsible for climate change.

I apply my fact-sensitive account of moral responsibility in which the fitness-conditions are normatively significant choice, sufficient knowledge, and control over the choice. Chapter 5 investigates the two first conditions about the choice and knowledge, and Chapter 6 examines the control condition. Subsequently, in Part 3, I show how these fitness-conditions feed into the discussion of who ought to be held morally responsible for climate change.

The tragedy of the few

The purpose of this chapter is to consider the first fitness-condition of responsibility – namely, to what extent institutional group agents face a *normatively significant policy choice* with regard to climate politics. In order to answer this question, we first have to consider what the theoretical understanding of the problem of climate change is. What types of problems are presumed to be politically challenging? As we saw earlier, the rational agent theory argues that the main challenge of climate politics today is *the tragedy of the commons*.

The tragedy of the commons is a suggested explanation of why depletion of natural resources is happening. This narrative argues that rational agents overuse the resources to which there is free access. This understanding of the problem fits well with its methodological and theoretical assumptions. However, as I argued earlier, when discussing the fitness-conditions of moral responsibility, it is fruitful to presume a different methodology. I find the most plausible methodology to be non-reductionist institutionalism. Within this framework, climate change and environmental harm cannot simply be explained by reference to an individual level of rational agency and self-interest theory alone.

By shifting the methodological level, a different layer of social agency and explanation is exposed. Here, overconsumption is not described and explained merely at the individual level, the institutional behavioural space of politics and institutions is considered as well. At this social and political institutional level, what explains overconsumption and climate change is not individual rational agents but the technical, social, legal, and political infrastructure. This approach differs from the earlier discussed Ostromian approach to institutions by not focusing solely on rational incentive modifications. Rational incentives may play a role but do not dominate the agents' acts and choices.

One social explanation that uses the political institutional approach addresses the modern technologies that make the overconsumption of natural resources possible. Another addresses the current legal and political landscape of policies that to a large extent allows a few multinational corporations to dominate global distribution and consumption of natural resources (Desombre and Barkin 2011, 50; Shearman and Smith 2007, Chapters 3 and 4). This dominance is achieved by various control mechanisms, such as quotas and subsidies for advanced technologies (IMF 2013). For example, this is the case in modern industrialised forms of fishing and farming in which highly effective and advanced technologies are subsidised by states and governments (Desombre and Barkin 2011, Chapter 2).[1] The same applies in the cases of carbon pollution from subsidised oil refineries and methane pollution from animal farming (Shearman and Smith 2007, Chapters 3 and 4). It is estimated that removing the current $480 billion worth of subsidies of petroleum products, electricity, natural gas, and coal would lead to a 13 percent decline in CO_2 emissions (IMF 2013, 1).

Moreover, Richard Heede argued in an article that 90 investor-owned and state-owned corporations are responsible for 63 percent of the cumulative worldwide emissions of industrial CO_2 and methane that occurred between 1751 and 2010 (Heede 2014, 229). By addressing these facts, climate change and carbon emissions are analysed in terms of the fossil fuels produced by incorporated entities, such as investor-owned or state-owned companies, rather than the result of the rational behaviour of consumers and individual polluters (Heede 2014, 231). Thus, the main social and political reason for overconsumption of the earth's natural resources is not that too many individuals overuse their fair share of natural resources, as suggested by *the tragedy of the commons* story. Rather, the current natural resource consumption is determined by a high level of industrial concentration and the fact that a few corporate agents are allowed to control the pollution of the earth's water, soil, air, and atmosphere (Desombre and Barkin 2011, 50; Shearman and Smith 2007, Chapters 3 and 4).

Once we recognise these facts, the challenge of environmental and climate sustainability becomes not merely a question of *free access* but rather one of *limited access* to the sustainable natural resources (Ostrom et al. 1994; Schlager et al. 1994; Sagoff 2008). I suggest we call this dilemma *the tragedy of the few*. Limited access to resources implies unevenly distributed access to the consumption and benefits of the resources. The reason for this is described in a recent report from the International Monetary Fund (IMF):

> While aimed at protecting consumers, subsidies aggravate fiscal imbalances, crowd-out priority public spending, and depress private investment, including in the energy sector. Subsidies also distort resource allocation by encouraging excessive energy consumption, artificially promoting capital-intensive industries, reducing incentives for investment in renewable energy, and accelerating the depletion of natural resources. Most subsidy benefits are captured by higher-income households, reinforcing inequality. Even future generations are affected through the damaging effects of increased energy consumption on global warming.
>
> (IMF 2013, 1)

By addressing this set of social and political facts about access and distribution, the content of climate politics is not merely a question of how to mitigate CO_2 emissions in a cost-efficient way, which was the core of the debate about the fitness of rational agents to be held responsible. Instead, other sets of questions should be addressed, including the social and political challenges that concern the distribution of benefits and the understanding of social and political conditions under which mitigation and adaptation policies are possible (Kaul and Mendoza 2002).

Note that by focusing on the benefits of natural resources instead of the costs of emission mitigation, the political management of climate change resembles the challenges related to other types of environmental harm. Conceived as a management problem of the collective benefits of natural resources, climate change is a type of pollution (Attfield 1999). Similarly, US president Barack Obama addressed this issue when he presented the *2013 Climate Action Plan* at Georgetown University. Climate change, Obama said, is not merely a challenge of long-term emissions and sharing costs and burdens. America should take national and global leadership in rectifying environmental harm and climate change by, among other things, regulating 'carbon pollution' in the same way other types of environmental pollution are being regulated with success (Obama 2013).

This is a reframing of the political challenge of climate change. Many perceive climate change as a 'perfect moral storm' (Gardiner 2011) and as a 'wicked problem' (Prins and Rayner 2007), indicating that climate change is very hard to manage politically. In contrast, by approaching the earth's climate as a resource and thus climate change as a challenge of carbon pollution, the problem of climate change becomes a politically manageable resource problem. Taking this reframing of the political challenge to be a reasonable and fruitful approach, the following discussion emphasises that the political challenge of environmental harm and climate change is one of avoiding further pollution and maintenance of sustainable natural resources.

As we shall see, this reframing of the political challenge of climate change is pivotal for the discussion of whether institutional group agents face a normatively significant choice with regard to providing effective policy solutions to environmental harm and climate change. The next sections lay out the normative and factual foundations for concluding whether the problem definition of *the tragedy of the few* allows for a normatively significant choice at the collective level of agency.

Natural resources and human rights

The tragedy of the few takes its point of departure from the fact that the earth's population has limited access to consumption and subtraction of resources, and that this limitation has morally significant constraints. The reason for this can be found in classic human rights theory. There are moral reasons to believe that all people have human rights to life, health, and subsistence (Caney 2010a; Shue 1980; Nussbaum and Sen 1993).[2] Equal access to resources and consumption marks a *moral threshold* below which people should not fall (Caney 2010a, 164). If access to resources is unevenly distributed, it is most likely that consumption and benefits will also be unevenly distributed.

Moral theories of climate change give us reason to criticise policies that promote unequal access to natural resources and to defend policies that ban environmental harm and pollution of water, air, soil, and atmosphere. This type of theoretical reasoning takes environmental harm and climate change to be morally wrong. The approach is comparable to other cases where concerns for, for example, human dignity have led to policies abolishing slavery, child labour, and other types of human rights infringement. Simon Caney defined the environmental rights in the following way:

> HR1, *the human right to life*: Every person has a human right not to be 'arbitrarily deprived of his life' (International Covenant on Civil and Political Rights, 1976, Article 6.1).
>
> (Caney 2010a, 166)
>
> HR2, *the human right to health*: All persons have a human right that other people do not act so as to create serious threats to their health.
>
> (Ibid., 167)
>
> H23, *the human right to subsistence*: All persons have a human right that other people do not act so as to deprive them of the means of subsistence.
>
> (Ibid., 168)

These rights can be considered as fundamental or basic rights that provide grounds for the impermissibility of depriving other persons their rights to life, health, and subsistence.[3] The implication is that it is impossible to trade some of these rights for other types of non-basic rights. Non-basic rights are defined as rights that can only be enjoyed if basic rights are secured. Thus, in light of *the tragedy of the few*, the political challenge is the question of *who* should be allowed to consume *how much* (Sagoff 2008, 59; Caney 2010a). I continue to further clarify the dilemma of *the tragedy of the few* by focusing on the physical properties of natural resources in the next sections. These factual circumstances are subsequently matched to different types of natural resource rights. Together with the normative considerations earlier, they feed into a fact-sensitive account of what policy options are available.

It is common in the economic literature to distinguish natural resources from public goods and private goods by means of two properties – namely, subtractability and exclusiveness (Ostrom 1990, 32; Ostrom 2003, 239; Kaul and Mendoza 2002, 83). This categorisation is appropriate in order to differentiate natural resources from private goods and public goods. Public goods and common-pool resources belong to the residual group of entities that fall outside the theory of market economy (Pindyck and Rubinfeld 2009).

Let us briefly distinguish the three different categories. (1) *Private goods* are the goods that can be traded on the market because they have excludable benefits and are rival in consumption (Kaul and Mendoza 2002 79). This does not apply to public and common goods. (2) *Public goods*, such as security, a well-functioning market, liberty democracy, and peace, cannot be traded on the market. One reason for this is that the benefits of public goods have non-excludable benefits and are non-rival in consumption, which means that the security of one person does not reduce or in any way deprive another person of the opportunity to consume the benefits of security. In contrast, (3) *common-pool resources* are subtractable and rival. Whereas peace cannot be over-consumed, common goods, such as fish stocks, water, air, and biocapacity, can be over-consumed and eventually depleted.

Both public and common goods are classified as market failures and as a result justify cases for government intervention. The market cannot price either public or common goods efficiently (Kaul and Mendoza 2002, 80; Sagoff 2008, 89). Non-market value is a characteristic that can be attributed to goods that achieve no value at the market but nonetheless are claimed to have a value. The values of a sustainable fish stock, clean air, peaceful cooperation, etc., are goods with non-market value. The reason why they achieve no value in the market is because they cannot be traded in marginal units (Sagoff 2008, 20).

The categorisation of private, public, and common goods, however, is too simple to understand the special social and political problems related to the dilemma of *the tragedy of the few*. Considerable progress has been made in the study of natural resources. One finding differentiates natural resources according to the physical properties of 'mobility flows' and 'storage'. Mobile resources are, for example, fish, and resources with storage properties are, for example, groundwater basins (Schlager et al. 1994, 296).[4]

Further points regarding the physical properties of natural resources should be added. First, not all types of natural resources are subtractable (Kaul and Mendoza 2002, 78). Nonsubtractable common-pool resources include natural resources that are non-rival, nonexclusive, and nonsubtractable. Good candidates for this type are sunlight and moonlight: the consumption of sunlight can neither be subtracted nor depleted. Nonsubtractable natural resources resemble the properties of public goods to which everybody has equal access.

Second, these kinds of natural resources are not private goods and cannot be bought or sold. Technically, we may utilise (and thereby subtract) the energy of sunlight in – for example, solar cells – but this utilisation does not in any way subtract the resources of sunlight. Moreover, using it does not limit others' consumption and benefit of sunlight. Sunlight cannot be transformed into tradable units because it is impossible to limit people's access to the good. In contrast, subtractive natural resources are scarce resources that can be depleted. It is important to note the fact that resources may shift between subgroups depending on the subtraction methods and channels that are available.

Whereas many natural resources in pre-modern times belonged to the group of nonsubtractable resources, they now belong to the subtractive group that can be depleted (Kaul and Mendoza 2002, 78). Take, for example, the case of the atmosphere. Until recently, the atmosphere and the stratosphere were nonsubtractable resources. Nonetheless, due to new technologies creating chemical changes in the environment (caused by pollution from chlorofluorocarbons, CO_2, and other gases), the atmosphere and the stratosphere have moved from the class of nonsubtractable resources to the class of subtractive resources (Kaul and Mendoza 2002, 78).

Two sets of rights

Let us now combine the insights on the factual properties of natural resources with the normative notions on human rights. This section demonstrates that there is a political challenge with regard to the clash between *flowing* natural resources and individual's territorial rights.[5] This discussion will allow us to conclude whether institutional group agents might face a normatively significant choice in regard to solving *the tragedy of the few*.

To clarify the challenge, the distinction between *territorially divided* and *non-divided* resources should be addressed. This distinction refers to what extent the consumption of the resource can be territorially limited or not. This relates to the question of what types of rights are relevant with regard to natural resources. If a farmer owns a piece of land, we normally do not mean that she has property rights over the worms in the soil or over the atmosphere above. Nonetheless, by utilising the land for farming, the farmer takes advantage of the pollination of crops by bees in the air and the transport of water and nutrients by worms and microbes in the soil. Due to territorial land rights, the farmer will be the only agent that has free access to the resources in the soil, water, and air on this particular piece of land. This is because the paradigm of property rights implies full ownership and control of access to the land surface. As we shall see in Chapter 8, there are valid liberal reasons for why people's negative rights and rights to private property ought not be infringed upon.

- To achieve a better understanding of the political challenges related to territorial rights and territorially non-divided resources, let us introduce a more detailed conceptualisation of rights. Property rights are frequently defined as equivalent to the right of excluding other people from one's territory (Ostrom 2003, 250). If a more nuanced categorisation of natural resources is applied, this concept of property rights is insufficient. In an article, Ostrom attached bundles of rights to different natural resources in order to differentiate between resources depending on the types of rights people are entitled to. She distinguished between rights resembling 'authorised entrant,' 'user', 'claimant', 'proprietor', and 'full owner' (Ostrom 2003, 251). If we accept this differentiation of rights, in many cases, the relevant property right to natural resources is not full ownership but a right to manage, use, or access. Let us look into why this is the case.

Take, for example, the farmers' utilisation of bees, worms, plankton, and other microorganisms in the production of agricultural products to which she has full ownership, which includes the right to sell. The farmer does not have full ownership over the bees, the worms, or the microorganisms in the soil. The relevant bundle of rights in such cases is, therefore, different from the right to, for example, the agricultural products. In the case of bees, worms, and microorganisms, we may say that the farmer has the *right to use* but *not full ownership*. This fits well with the fact that these animals and resources are territorially non-divided natural resources and are thus difficult to subject to private ownership. The same challenge applies to different kinds of natural resources, such as the ability of fertile soil to grow crops, photosynthesis, air, and wind (Sagoff 2008, 103). In contrast, subtractive and territorially divided natural resources, such as oil, gas, coal, minerals, and agricultural products, are normally resource units that are subjected to either full private ownership or proprietor rights.

Conceived this way, *the tragedy of the few* suggests an alternative explanation of why environmental harm and climate change happens that departs from the standard account of *the tragedy of the commons*. The reason for this departure is that *owners of land and technological equipment have a privileged right to subtract and use resources to which the owners have no full ownership*. Arguably, the political challenge is how to balance the two sets of abstract human rights to life, health, and subsistence on the one hand, and territorial rights on the other. Assuming that *the tragedy of the few* captures the dilemmas in relation to environmental harm and climate change, we are also interested in the question whether the relevant institutional group agents are faced with normatively significant policy options. Is it possible to balance the two sets of rights in a normatively consistent manner?

The policy of access

One way to approach the political challenge of balancing rights to resources and territorial rights is to provide a fact-sensitive and concrete account of the relevant rights. As I argued in Part 1, abstract normative principles fail to guide political regulation due to the indeterminacy challenge. These principles say too little about the factual component of the challenge. The essential question is, therefore, what the rights would look like in a fact-sensitive interpretation. If the factual circumstances of natural resources are combined with the normative concerns for access and consumption rights to uncontaminated natural resources, we can suggest a more precise account of the clash between bundles of territorial rights and people's human rights to life, health, and subsistence.

In light of the current challenges of (1) depletion of resources and (2) limited territorial owner or proprietor rights, the basic and fundamental human rights to life, health, and subsistence can be re-formulated as concrete basic and non-basic rights. Good candidates for concrete basic rights are *equal right to access* resources and consumption of sustainable resources (Kaul and Mendoza 2002, 92). On the other hand, good candidates of concrete non-basic rights are the right to subtract resource units or alternatively, the right to emit carbon. Systematising the discussion of the norms and facts related to *the tragedy of the few*, the following normative and factual circumstances of the earth's natural resources should be addressed:

1   *The norms of natural resources*: the rights to access and consume vital natural resources; and
2   *The facts of natural resources*:

    a   the physical properties of the resource;
    b   the necessity of human consumption and access; and
    c   the degree of potential depletion.

Based on these considerations, four normatively significant policy designs can be differentiated:

- *No regulation/restriction*: if the resource units are not subtractive, such as sunlight, access to and consumption of the resource is neither a moral nor political challenge. Thus, the use should not be restricted.
- *Distribution of access*: if the resource units are (1) subtractive and (2) covered by people's basic rights but (3) territorially limited, access to and consumption of the resource is a moral and political challenge. Thus, a political design is required in order to protect people's equal right to access and consumption by distributing access to relevant territories with resources nationally and globally.
- *Political regulation/ban*: if the resource units are (1) subtractive and (2) covered by people's basic rights but (3) territorially unlimited, access to and consumption of the resource is a moral and political challenge. Thus, a sophisticated political design is needed that limits the overuse of these resources and safeguards resource maintenance and equal opportunity for future generations to access and consume.
- *New technologies*: if new technologies and subtraction methods are developed, new political designs should be considered.

(Schomberg 2011, 20)

To consider whether these normatively significant policy choices are available in the methodology of the institutional group agents, one last point should be considered. According to the non-reductionist methodology of the institutional group agents, the individual level of agency sustains its moral attribution. One implication of this is that policy options relevant to the group agents must not infringe on individuals' freedom and territorial rights. This point relates to the discussion of positive and negative rights, to which I return in Chapter 8. People's negative rights are not allowed to be infringed upon because we have a moral reason not to physically force people to do something that they do not voluntarily accept.

The vital question is, do the four policy designs infringe upon individuals' freedom and negative rights? Differentiating the different bundles of rights provides reasons for why this is not the case. Let us explore this. Policies for natural resources have the potential to infringe people's negative rights and their ownership rights because the protection of natural resources clashes with individuals' territorial rights, as we saw in the previous section. On the other hand, individuals with territorial rights have no ownership rights to the bees, the rain worms, the microorganisms, and clean atmosphere, air, water, and soil. Owners with territorial rights have a *right to use* and thereby take advantage of the resources in the atmosphere, air, water, and soil. But they do not have full ownership rights, and thus no *right to deplete, exploit*, or *pollute*.

Hence, political regulations of natural resources cannot be an infringement of full ownership and proprietor rights. Rather, a political regulation or ban ensures that people do not misuse their territorially or technologically privileged rights to exploit common-pool resources in the atmosphere, air, water, and soil, and, moreover, do not deprive other people and future generations of the ability to access and consume these resources.[6] Given that people's rights to access, consumption, and subtraction are warranted by basic human rights to life, health, and subsistence, these basic rights should be politically prioritised over non-basic rights to use. The objectives of environmental and climate politics are to secure people's rights to life, health, and subsistence by limiting carbon pollution, waste disposal, and chemical products and to protect nature in order to secure people's basic rights to natural resources. With this in mind, we can conclude that institutional group agents face several normatively significant policy options with regard to the problems of *the tragedy of the few*. The policy options are compatible with the methodological assumptions of institutional group agency theory and normative theory for individual liberal rights.

Access to knowledge

This discussion provides a logical argument that the first condition for holding institutional group agents collectively responsible for climate change is satisfied. Institutional group agents may be fit to be held responsible for climate change because they have several normatively significant *options of access* to resources and distribution of benefits. In the following sections, I consider the second condition – i.e., in order to hold institutional group agents morally responsible, they should have access to the evidence required for making normative judgements about their policy options (List and Pettit 2011, 158). Two institutional designs for knowledge sharing are examined. If they are successful, they support the thesis that institutional group agents are fit be to held responsible for climate change. It should be emphasised that the methodological assumptions allow both for individualistic and for socially and institutionally mediated knowledge building. First, we consider the social institution of the market and then the political institution of democracy.

I argued earlier that the rational-individualist approach presumes a methodology to agency, rationality, and politics that understands climate change as a problem of incentives and motivations. More specifically, Chapter 4 showed that self-interest theory is individually and collectively self-defeating and that it is a mistake to suggest self-interest theory as the only explanatory model for climate change. I also argued that self-interest theory cannot explain why rational agents underestimate the collective benefits of environmental protection because the explanation presumes what it wants to explain – namely, that rational agents underestimate collective benefits. The next section discusses whether the problems of rational preference-making may evaporate at the institutional and group level of the market. The discussion concerns whether the social institutions of the market eliminate the ignorance about collective benefits, which was identified earlier as the core challenge of *the tragedy of the commons*.

The market and values

Friedrich Hayek is famous for having defended the market as an efficient distributor of knowledge among rational agents. According to Hayek, modern societies face a knowledge problem:

> The peculiar character of the problem of a rational economic order is determined precisely by the fact that the knowledge of the circumstances of which we must make use never exists in concentrated or integrated form, but solely as the dispersed bits of incomplete and frequently contradictory knowledge which all the separate individuals possess.
>
> (Hayek 1945, 519)[7]

Hayek's account of the knowledge problems has two functions. First, it explains why political macro-level planning is impossible. Individuals and thus also political decision making suffer from incomplete knowledge about the value of goods, the total number of activities in society, and what would be the right decision at the right time. Thus, political (economic) planning is impossible because individuals who engage in political planning are unable to estimate the value of goods in the same way, as we saw earlier, because they are unable to estimate the value of collective benefits (Hayek 1945, 524).[8]

Two premises must be accepted in order to embrace this argument. First, politics is neither an aggregate of individual people's knowledge nor a knowledge-producing process. Hayek envisages a market capable of facilitating and providing value judgements and decisions without a centre of political planning. Note that this critique of political decision making is not a critique of political coordination of societal behaviour as such: it is a critique of the centralisation of political coordination. Another way of criticising the centralisation of politics is an argument for bottom-up, disaggregated, and polycentric political decision-making processes that are open to critique, comments, and discussions (Prins and Rayner 2007; Stehr and Storch 2009; Ostrom 2012; Stehr 2013; Rayner 2013). Here, disaggregating decision-making power solves the valuation and deliberation errors of centralised political decision processes, to which I return later.

The second function of the knowledge problem is the mechanism that drives activities in the market. Because people – e.g., individual producers and consumers – do not have relevant information about market products at their disposal, they refer to the market as a way of dispersing knowledge about the value of certain goods and products. The market conveys this information by price competition, which operates as a 'discovery procedure' that secures the utilisation of disaggregated knowledge (Hayek 1973, 117). Hayek explains the social dynamics of knowledge diffusion between agents in the market in the following way:

> The chief cause of the wealth-creating character of the game is that the returns of the efforts of each player act as the signs which enable him to contribute to the satisfaction of needs of which he does not know, and to do so by taking advantage of conditions of which he also learns only indirectly through their being reflected in the prices of the factors of production which they use. It is thus a wealth-producing game because it supplies to each player information which enables him to provide for needs of which he has no direct knowledge and by the use of means of the existence of which without it he would have no cognizance, thus bringing about the satisfaction of a greater range of needs than would otherwise be possible.
>
> (Hayek 1973, 115)

The market coordinates and facilitates knowledge dispersion and thereby corrects the valuation and deliberation errors conducted by individual agents. Because the market conveys value judgements in prices, the challenges of the knowledge problem are eliminated. Before we accept this conclusion, however, the compatibility between the concept of value related to collective benefits ignorance and the concept of value related to Hayek's knowledge theory must be examined. Of particular importance is whether the social institutions of the market can eliminate collective benefits ignorance at the individual level of agency. In order to comprehend whether the market balances individuals' collective benefits ignorance, an important question is whether the market helps individual agents create *subjective* values or to acknowledge the *objective* value of goods.[9]

Two distinctions should be added. The first distinction concerns whether the *source* of values is objective or subjective. A subjective valuing source is an agent who values an object where the values depend on the attitudes and opinions of the particular individual.[10] An objective valuing source is when the value of an object is independent of the human value. Objective valuation requires a normative theory to justify the objective value of an object.

The second distinction concerns how to comprehend the objective value of entities, such as natural resources. John Searle distinguishes what is *epistemically objective* and *subjective* from what is *ontologically objective* and *subjective*. By *ontologically subjective*, Searle means that institutions and norms are social constructs that would not exist if human agency did not exist. Institutional and social facts do not constitute an independent ontology but nonetheless represent an epistemic objective reality that individuals collectively share (Searle 2010, 17–18).

According to Searle, this is possible because human beings have (1) the capacity to share ideas, norms, beliefs, and intentions, and (2) the capacity to assign functions to social entities. Take the example of a government. It is a social institution: it consists of buildings and groups of people organised around a set of practices. However, the buildings and the groups of people alone are not a sufficient description of the government as a social institution with legitimate authority. The defining component of a government is not its physical organisation but the social attitudes towards the government. Here, social attitudes are taken to designate, for example, collective experiences and intentions shared by citizens and other agents. The institutional level of agency, social attitudes, and shared social facts depend on collective experiences and qualities that allow for institutions and norms to be *epistemically objective* (i.e., shared among many people).

Because the market appreciates this way of understanding social norms and the earlier developed account of the supervenience relationship between the collective and individual level of conduct, it can be perceived as not merely a venue for exchange of individuals' rational and subjective preferences (Ray 2009, Polanyi 1974). In contrast, the market, prices, and values are social institutions and norms that are ontologically subjective but epistemically objective (Searle 2010, Chapter 5). When we consider the market as a social phenomenon, the prices of entities and resources do not express the individual agent's subjective valuing of an entity. Instead, the price system is a social conductor of communication, knowledge, and value.[11]

The market and collective benefits ignorance

In the previous section, we considered to what extent the theory of collective benefits ignorance and Hayek's knowledge theory are compatible by focusing on whether the market conveys the knowledge necessary for compensating for collective benefits ignorance with regard to environmental sustainability. To further elaborate this argument, let us consider Hayek's example of a purchase of aluminium instead of magnesium, which exposes how the function of competition and pricing at the market approximates commodities' objective value but not their subjective value. Hayek envisioned a producer that one day buys aluminium instead of magnesium because the price of the former has fallen relative to that of the latter (Hayek 1973, 116). Because the producer does not have a subjective preference for aluminium or magnesium, the price of the two materials cannot reflect the producer's subjective values. Instead, the market prices reveal what Hayek called 'the interests of the community at large' (Hayek 1973, 116). When the producer buys aluminium instead of magnesium, he is not acting according to his own private value assessments but conforms to what the market communicates is a rational and valuable choice. Hayek explained,

> [. . .] in the market order each is made by the visible gain to himself to serve needs which to him are invisible, and in order to do so to avail himself of to him unknown particular circumstances which put him in the position to satisfy these needs at as small cost as possible in terms of other things which it is possible to produce instead.
>
> (Hayek 1973, 116)

This quote, together with the one cited earlier, gives us reason to believe that the prices of goods do not reflect subjective value assessments. In contrast, the market is a social institution that is a conductor and distributor of social knowledge and values by providing the mechanism for how to estimate the epistemically *objective* value of a specific good that in other societal contexts would be tacit to the producer and consumer.[12] So, Hayek's understanding of value judgement does not presume a subjective theory of value but a social theory of objective values.

If this is correct, Hayek's market theory does not solve the self-interest problems of collective benefits at the individual level. As I argued in Chapter 4, the self-interest theory presumes an individualist methodology which rejects social norms and institutional facts at the collective level of agency. According to the individualist, the sources of pricing and valuing are subjective. The self-interest problems with regard to calculating the benefits of environmental sustainability are thus unfit to be solved at the market. This is because market theory relies on an epistemology that is incompatible with the reductionist rational version of individualism. Note that this does not imply that the account of the market institutions is incompatible with a political-normative theory of individualism and liberalism.[13]

Having explored in general terms the relationship between the market and valuing of objects, let us now turn to what implications this might have for the agents to obtain knowledge about environmental and climate matters through the market. Can the market reveal the 'objective' value of natural resources? As we have seen, Hayek used an epistemic argument to introduce the market competition because the market is an efficient way to distribute knowledge about private commodities whose values can be approximated by exchange value and pricing (Pleasants 1997; O'Neill 2006).

Note that this does not imply that all kinds of values should be determined by the market, nor that individuals cannot obtain knowledge about, for example, environmental values through other social institutions. If the market is a favourable institution that reveals and collects knowledge that the individuals did not have prior to exchange at the market, the implication is that the values individual agents hold do not need to be valued at the market. I suggest that this applies to environmental sustainability. Hayek did not say much about environmental values. At one point in his book, *The Road to Serfdom*, he addressed negative societal effects on the environment, recognising that natural resources and other public goods cannot be captured as private goods tradable at the market:

> In all these instances there is a divergence between the items which enter into private calculation and those which affect social welfare; and whenever this divergence becomes important some method other than competition may have to be found to supply the services in question. Thus neither the provision of signposts on the roads, nor, in most circumstances, that of the roads themselves, can be paid for by every individual user. Nor can certain harmful effects of deforestation, or of some methods of farming, or of the smoke and noise of factories, be confined to the owner of the property in question or to those who are willing to submit to the damage for an agreed compensation. In such instances we must find some substitute for the regulation by the price mechanism.
>
> (Hayek 1944, 40)

In this passage, Hayek argued that not all policy areas should be submitted to market competition and that the value and pricing mechanism of the market is suitable for some types of goods. This resembles the distinction between private goods and public goods, and common-pool resources that were outlined earlier. The market is not a productive social institution for evaluating environmental harm, deforestation, and other types of 'common bads'. Knowledge and values about environmental and deforestation need other social sources.

To summarise, Hayek's account of the market cannot solve the collective benefits ignorance at the individual level of conduct. His ideas of dissemination of tacit knowledge through the market mechanism are suitable for distributing knowledge about private goods but not goods and commodities that individuals value independently of the market (i.e., public goods and common-pool resources). Moreover, if we allow for individuals and groups to value the sustainability of the earth's environment and climate in the first place, we do not need a market theory that provides and distributes knowledge about the value of the earth's environment and climate. In contrast, as I argue in Chapter 9, it defeats the purpose of exposing entities, such as natural resources, for market competition in order to estimate the value thereof. Instead, I argue that we should be occupied with how to establish political institutions through which an agent's green values can be realised.

Based on this, we can conclude that it is not unequivocally clear that the second fitness-condition of moral responsibility with regard to access to knowledge is satisfied. There are considerable doubts about the market theory as an institution that provides the necessary knowledge and evidence for the normative, social, and physical properties of natural resources that should feed into relevant environmental and climate policy options. In Chapter 6, I consider whether Hayek is useful in the current discussion of market-based climate solutions.

Democratic climate governance

Do democratic institutions provide an institutional design that satisfies the second fitness-condition concerning sufficient knowledge about environmental harm and climate change? The discussion of whether democratic political institutions provide a more suitable framework than the market does not take its point of departure from the discussion on collective benefits ignorance. The democratic theory discussed here takes a different point of view by focusing on how free education, science, and deliberation allow for knowledge-producing social engagement (Rostbøll 2009; Ansell 2012). I elaborate in this section on the conceptualisation of why deliberation within democratic institutions enhances the social level of knowledge. My argument is epistemic: knowledge becomes statistically more robust in institutional designs that are capable of producing, enhancing, and collecting knowledge (Popper 2002; List and Pettit 2011).

Open and free societies are better at avoiding mistakes and misinformation by increasing the number of free and competent people engaged in scientific and non-scientific evaluations and assessments (List and Pettit 2011, 90). This is not an uncontested argument: many scholars have criticised the knowledge sharing problems of democracies.[14] Notwithstanding this, I find the argument that democracies can be more robust in producing and collecting knowledge sustains some plausibility, and continue to elaborate on this characteristic.

In my view, democratic majority voting and deliberation outperforms both dictatorial and unanimity rules in terms of maximising the group's reliability on a given proposition. Note that the premise for this is that people are (1) independent, (2) fallible, and (3) biased towards the truth in their judgement (List and Pettit 2011, 90). This means that agents participating in the vote and deliberation should be allowed to vote and express their opinions independent of other agents, and moreover, that agents can make fallible assertions that nonetheless are assumed to be biased towards the truth.

To be more precise: when a group of independent and competent people seeks to track the truth, democratisation of the process increases the epistemic gains of the knowledge-seeking process (List and Pettit 2011, 90). Consider one example. Scientist Francis Galton investigated a contest in which 800 participants were asked to estimate the weight of an ox who was on display. Whilst few participants were individually accurate, the average estimate across the group of participants turned out to be 1197 pounds, almost identical to the ox's true weight of 1,198 pounds (List and Pettit 2011, 86).

There is a statistical explanation for why there is an epistemic gain from taking the average of a large number of independent and fallible judgements. The point is that the likelihood that *one* judgement is accurate is very low. In contrast, the likelihood is very high that the *average* of all judgements is accurate. Moreover, accuracy increases if the contest is repeated, taking the average of a new collection of judgements. In other words, the average of the average has a very high likelihood of being very accurate. This type of information pooling is harnessed in democracies in which each citizen is allowed to vote in ballots and referendums, participate in public debates, and influence politics in other ways.

Further, if the conditions for information pooling in democratic institutions are the independence and competence of the participants (List and Pettit 2011, 87), we may derive the following notion: that threats to democratic politics of climate change are illiteracy, underinvestment in prime schooling, and the lack of free access to scientific results and deliberation in public media. People are only competent and independent to engage in democratic climate governance if their access to free media, education, and science is not limited.

Based on this, a theoretical correlation can be suggested between the democratic institutional design of free media, education, and science on the one hand and a society's ability to address climate change politically on the other. In other words, how well a democratic society can address climate change correlates with how well the organisational structure can make use of the aggregated epistemic gains that people collectively propel. According to this theory, centralisation of the political process will decrease the epistemic gains.

This claim resembles Hayek's critique of centralised political planning. In the same way as the market benefits from independent and competent participants with regard to valuing market goods, political democracies benefit from independent and competent citizens with regard to tracking the best policy options. Note, however, that the two types of aggregation and collection of information differ fundamentally. Whilst the collection of market information is based on a fixed understanding that competition and economic incentives drive people's choices, democratic information pooling is based on independence, competence, and bias towards the truth.

Conclusion

In this chapter, I have applied the fact-sensitive model of moral responsibility to the non-reductionist institutional methodology by focusing particularly on the first and second conditions of fitness. I have argued that the methodology of non-reductionist institutionalism provides a framework within which group agents are fit to be held morally responsible with regard to the first and second conditions.

I have also argued that the political problems of climate change are related to the questions of equal access and benefits distribution related to *the tragedy of the few*. I have argued that democratic political institutions provide the best structure for providing the necessary information about climate governance. In contrast, I have demonstrated that the social institutions of the marketplace cannot remedy the collective benefits ignorance that rational agents are expected to exhibit.

The next chapter discusses the third fitness-condition about control over the morally significant policy options. Although the first condition of responsibility is satisfied in the sense that institutionalised group agents face a normatively significant set of policy choices, and the second condition is satisfied in the sense that democratic group agents allow for an effective knowledge pooling, we may still reject that democratic group agents are fit to be held responsible if their institutional structure is not strong enough to control the relevant policy options.

Notes

1 Political regulation has subsidised the modernisation of fishing fleets, which has led to a disproportionately efficient fishing industry. Regulations designed by WTO and the European Union contribute to limiting the access to the subtraction channels by subsidising the most effective and highly technological fishermen (Desombre and Barkin 2011, Chapters 3 and 4).

2 Much attention has been paid to the political and civil aspects of human rights. Less attention has been paid to the environmental and subsistence aspects of human rights. Important exemptions are Shue's *Basic Rights* (1980), Martha Nussbaum's and Amartya Sen's edited volume on *The Quality of Life* (Nussbaum and Sen 1993) and more recently Simon Caney's 'Climate Change, Human Rights and Moral Thresholds' (2010a).

3 Following Shue, rights are basic 'if enjoyment of them is essential to the enjoyment of all other rights. This is what is distinctive about a basic right. When a right is genuinely basic, any attempt to enjoy any other right by scarifying the basic right would be quite literally self-defeating, cutting the ground from beneath. Therefore, if a right is basic, other, non-basic rights may be sacrificed, if necessary, in order to secure the basic right' (Shue 1980, 19).

4 Moreover, Schlager et al. argued that the different sets of physical properties influence individual agents' incentives and strategies: 'The two characteristics of mobility and storage affect (1) the severity of [allocation and development problems], (2) the relative ease with which users can resolve those problems, and (3) the kinds of institutional arrangements they are likely to development and implement' (Schlager et al. 1994, 296–297).

5 Considerations of the territorial rights of nation states are here set aside. For a recent contribution, see Stilz (2009).

6 This conclusion is compatible with the left-libertarian account of 'joint ownership' and 'world ownership' of natural resources. Steiner formulates the essential point as follows: 'It is a necessary truth that no object can be made from nothing, and hence that all titles to manufactured or freely transferred objects must derive from titles to natural and previously unowned objects' (Steiner 1977, 44). For discussions of the idea of left-libertarianism, see Cohen (1995, Chapter 3) and Otsuka (2003). For a critical discussion, see Arneson (2010).

7 For Hayek, the knowledge problem is a distinctive feature of advanced modern civilisations: '[T]he fact that each moves within a coherent structure most of whose determinants are unknown to [each member of the society]. [. . .] The incurable ignorance of everyone [. . .] is the ignorance of particular facts which are or will become known to somebody and thereby affect the whole structure of society. This structure of human activities constantly adapts itself, and functions through adapting itself, to millions of facts which in their entirety are not known to anybody' (Hayek 1973, 13–14).

8 For a discussion of Hayek's epistemological critique of socialism, see Pleasants (1997) and O'Neill (2006).

9 For an elaborated account of values, see Parfit (2011, Chapters 2 and 3).

10 The subjective source does not imply a subjectivist – i.e., relativist – understanding of an object's values. The (subjective) valuer is capable of attributing both intrinsic and instrumental value to objects. Furthermore, there are two variants of instrumental

value. One assigns instrumental value to one's own desires and interests, and the other assigns value to others' desires and interests. Another way to frame the difference is to distinguish between egoistic and altruistic preferences. Both variants are consistent with consumer preference theory and social choice theory. Examples of altruistic or green consumer preferences are fair-trade products and certified green products. These products have internalised the costs of all production steps. The result is that the income from selling the product is distributed in order to protect, among other things, workers' rights and sustainable resource usage. The majority of the world's commercial goods is not produced or sold under these conditions (Elliott 2012; Mont et al. 2013).

11 To further substantiate this argument, we can supplement this understanding with some of Popper's observations. In the *Logic of Scientific Discovery* (1959), Popper argued that scientific knowledge is objective but at the same time fallible and thus provisional. This means that knowledge can be 'inter-subjectively tested' by using accepted methods and theories. Knowledge is, in other words, testable and hence open for refutation and falsification (Popper 1959, 22; see also Rostbøll 2009).

12 For a classic account of tacit knowledge, see Polanyi (1967). For a recent discussion of tacit knowledge in the work of Polanyi and Hayek, see, e.g., Perraton and Tarrant (2007).

13 Indeed, Hayek is proponent of a strong non-institutionalised individualism.

14 For a good overview of this discussion, see List and Goodin (2001). In the next chapter, I address several of these critics specific to democratic climate governance.

References

Ansell, C. K., 2012. *Pragmatist democracy: evolutionary learning as public philosophy*. Oxford: Oxford University Press.

Arneson, R. J., 2010. Self-ownership and world ownership: against left-libertarianism. *Social Philosophy and Policy*, 27 (Winter), 168–194.

Attfield, R., 1999. *The ethics of the global environment*. Edinburgh: Edinburgh University Press.

Caney, S., 2010a. Climate change, human rights and moral thresholds. *In:* S. Humphreys, ed. *Human rights and climate change*. New York, NY: Cambridge University Press, 69–91. Reprinted in 2010: *In:* S. M. Gardiner, S. Caney, D. Jamieson, and H. Shue, eds. *Climate ethics: essential readings*. Oxford: Oxford University Press, 163–181.

Cohen, G. A., 1995. *Self-ownership, freedom, and equality*. Cambridge: Cambridge University Press.

Desombre, E. R. and Barkin, J. S., 2011. *Fish*. Cambridge: Polity Press.

Elliott, K., 2012. *Is my fair trade coffee really fair? Trends and challenges in fair trade certification*. CGD Policy Paper 017. Washington, DC: Center for Global Development. Available from: www.cgdev.org/content/publications/detail /1426831 [Accessed 3 March 2016].

Gardiner, S. M., 2011. *A perfect moral storm: the ethical tragedy of climate change*. Oxford: Oxford University Press.

Hayek, F., 1944. *The road to serfdom*. New York, NY: Routledge.

Hayek, F., 1945. The use of knowledge in society. *The American Economic Review*, 35 (4), 519–530.

Hayek, F., 1973. *Law, legislation and liberty, Volume 1: rules and order*. Chicago, IL: The University of Chicago Press.

Heede, R., 2014. Tracing anthropogenic carbon dioxide and methane emissions to fossil fuel and cement producers, 1854–2010. *Climatic Change*, 122 (1–2), 229–241.

IMF (International Monetary Fund), 2013. *Energy subsidy reform: lessons and implications*. Available from: www.imf.org/external/np/pp/eng/2013/012813.pdf [Accessed 1 April 2016].

Kaul, I. and Mendoza, R. U., 2002. Advancing the concept of public goods. *In:* R. A. Musgrave and P. B. Musgrave, eds. *Providing global public goods*. Oxford: Oxford University Press, 125–143.

List, C. and Goodin, R. E., 2001. Epistemic democracy: generalizing the condorcet jury theorem. *Journal of Political Philosophy*, 9 (3), 277–306.

List, C. and Pettit, P., 2011. *Group agency: the possibility, design, and status of corporate agents*. Oxford: Oxford University Press.

Mont, O., Heiskanen, E., Power, K., and Kuusi, H., 2013. *Improving nordic policymaking by dispelling myths on sustainable consumption*. Copenhagen: Nordic Council of Ministers.

Nussbaum, M. C. and Sen, A., eds., 1993. *The quality of life*. Oxford: Clarendon Press.

Obama, B., 2013. *Remarks by the President on climate change*. Georgetown University, Washington, DC, 25 June, 1–7.

O'Neill, J., 2006. Knowledge, planning, and markets: a missing chapter in the socialist calculation debates. *Economics and Philosophy*, 22 (1), 55–78.

Ostrom, E., 1990. *Governing the commons: the evolution of institutions for collective action*. Cambridge: Cambridge University Press.

Ostrom, E., 2003. How types of goods and property rights jointly affect collective action. *Journal of Theoretical Politics*, 15 (3), 239–270.

Ostrom, E., 2012. Green from the grassroots. *Project Syndicate*, 12, 1–3.

Ostrom, E., Gardner, R., and Walker, J., 1994. *Rules, games, and common-pool resources*. Ann Arbor, MI: University of Michigan Press.

Otsuka, M., 2003. *Libertarianism without inequality*. Oxford: Oxford University Press.

Parfit, D., 2011. *On what matters*. Oxford: Oxford University Press.

Perraton, J. and Tarrant, I., 2007. What does tacit knowledge actually explain? *Journal of Economic Methodology*, 14 (3), 353–370.

Pindyck, R. S. and Rubinfeld, D. L., 2009. *Microeconomics*. 6th edn. International. Upper Saddle River, NJ: Pearson Prentice Hall.

Pleasants, N., 1997. The epistemological argument against socialism: a Wittgensteinian critique of Hayek and Giddens. *Inquiry*, 40 (1), 23–45.

Polanyi, M., 1967. *The tacit dimension*. London: Routledge.

Polanyi, M., 1974. *Personal knowledge: towards a post-critical philosophy*. Chicago, IL: University of Chicago Press.

Popper, K., 1959. *Logic of scientific discovery*. 1st English edn. New York, NY: Routledge. *In:* German: Popper, K., 1935. *Logik der forschung*. Vienna: Verlag von Julius Springer.

Popper, K., 2002. *The poverty of historicism*. Berkshire: Routledge.

Prins, G. and Rayner, S., 2007. *The wrong trousers: radically rethinking climate policy*. Available from: eureka.sbs.ox.ac.uk/66/ [Accessed 26 August 2017].

Ray, T., 2009. Rethinking Polanyi's concept of tacit knowledge: from personal knowing to imagined institutions. *Minerva*, 47 (1), 75–92.

Rayner, S., 2013. *Polycentric policy architectures*. Presented at Copenhagen Climate Workshop Series, 10 June 2013. Department of Political Science and Sustainability Science Centre, University of Copenhagen (unpublished).

Rostbøll, C., 2009. *Deliberative freedom: deliberative democracy as critical theory*. Albany, NY: State University of New York Press.

Sagoff, M., 2008. *The Economy of the earth, philosophy, law and the environment*. New York, NY: Cambridge University Press.

Schlager, E., Blomquist, W., and Yan Tang, S., 1994. Mobile flows, storage, and self-organized institutions for governing common-pool resources. *Land Economics*, 70 (3), 294–317.

Schomberg, R., 2011. Prospects for technology assessment in a framework of responsible research and innovation. *In:* M. Dusseldorp and R. Beecroft, eds. *Technikfolgen abschätzen lehren: Bildungspotenziale transdisziplinärer Methoden*. Wiesbaden: Vs Verlag.

Searle, J., 2010. *Making the social world: the structure of human civilization*. New York, NY: Oxford University Press.

Shearman, D. and Smith, J. W., 2007. *The climate change challenge and the failure of democracy*. Westport, CT: Praeger.

Shue, H., 1980. *Basic rights: subsistence, affluence, and U.S. foreign policy*. 2nd edn. Princeton, NJ: Princeton University Press.

Stehr, N., 2013. An inconvenient democracy: knowledge and climate change. *Society*, 50, 55–60.

Stehr, N. and Von Storch, H., 2009. Climate protection. *Journal of Consumer Protection and Food Safety*, 4, 56–60.

Steiner, H., 1977. The natural right to means of production. *Philosophical Quarterly*, 27 (106), 41–49.

Stilz, A., 2009. Why do states have territorial rights? *International Theory*, 1 (2), 185–213.

UN General Assembly, International Covenant on Civil and Political Rights, 16 December 1966, United Nations, Treaty Series, vol. 999, p. 171, available at: https://www.refworld.org/docid/3ae6b3aa0.html [accessed 12 July 2019].

6 Control conditions and democratic climate governance

As noted in the introduction, inquiries of fitness-conditions for responsibility for climate change touch upon the current critique of the strength of democratic governance (Held and Hervey 2009; Persson and Savulescu 2012). Although democratic institutions provide a framework within which individuals and groups can govern climate change, there is an increasing concern that democratic institutions do not *control* how climate change is managed politically. Control as a fitness-condition for moral responsibility concerns whether the relevant group agent has control over the morally significant choices and relevant policy options.

This critique questions the capability of democratic decision makers to develop policies and laws suitable for managing complex matters such as climate change. However, keep in mind that the critique is potentially misleading because, as we saw earlier, there are conceptual differences between how the political problems of climate change are addressed. In the rational account, the main problem is the economic incentives problems related to *the tragedy of the commons*; in the institutional account, particularly on the fact-sensitive notion defended in this book, the main problem is the distribution of access and benefits related to *the tragedy of the few*.

In this chapter, I consider the critique of democratic climate governance and note that it is not a normative critique of democratic ideals but a critique of democratic institutions' lack of *control* with satisfactory climate politics. My concern is that the critique of democratic control with climate governance presumes the political problems of climate governance to be the economic incentive problems of *the tragedy of the commons*.

Consider one of the most popular lines of argument: let us call it the *incentive critique*. The point of departure for the *incentive critique* is the scientific fact that the earth's climate is changing, possibly with catastrophic impacts (IPCC 2013) and that current democracies have failed to govern climate change satisfactorily. The critique's main argument is that political-democratic institutions are ill-suited for responding to climate change with sufficient means and that liberal democracies are inherently reluctant to mitigate environmental harm and climate change (Mathews 1991; Shearman and Smith 2007).

There are a number of reasons why this might be the case. For one thing, human beings might not have 'the nature equipped with a moral psychology that empowers them to cope with the moral problems that these new conditions of life create' (Persson and Savulescu 2012, 1); for another, the currently favoured political system of liberal democracy cannot overcome these moral deficiencies. In fact, this critique argues, liberal democracy makes these problems more acute (Shearman and Smith 2007, 165; Lovelock 2009; Persson and Savulescu 2012, 1). This is primarily because citizens, corporate stakeholders, and politicians do not have the necessary motivation or economic incentives to do so. Thus, the incentive critique concludes that democratic institutions fail to control the normatively significant choices and policy options that are present in the issue of climate change.

In order to overcome this challenge, some scholars argue that the rebuilding of political climate control requires that incentives be restored as a key component of climate mitigation. The incentive critique proposes two, albeit very different, solutions to the control the deficit of democratic climate governance. One way to change people's motivation and incentives is to use the price mechanism available in the marketplace. I call this strategy *neoliberal policy proposals*. Another way is to promote proactionary pharmacological changes in people's motivations and incentives, which we can call *proactionary policy proposals*. Now, the pertinent question is whether these two sets of policy proposals provide a stronger account of political control of climate governance than the democratic model.

The neoliberal policy proposals

There is a growing consensus that policies that utilise market mechanisms and economic incentives can be characterised by the term *neoliberalism* (Plant 2009; Peck 2010; Amable 2011). Saad-Filho and Johnston argued that neoliberalism is 'the dominant ideology shaping our world today' and that 'we live in the age of neoliberalism' (Saad-Filho and Johnston 2005, 1).[1] Common for the proponents of neoliberalism is their distrust of state-centred governance and management, which emerged as a reaction to the political-economic planning of fascist and socialist states in the 1920s–1970s (Fukuyama 2004, 119).[2]

The critique of the big state and state-centred economic policies has led to privatisation, liberalisation, and market-based policies in the past two-three decades (Harvey 2007; Buch-Hansen and Wigger 2011). These policy reforms steadily spread to the rest of Europe and still remain dominant in the contemporary politics of the US, Europe, and Asia. Moreover, the neoliberal approach dominates the politics of the major international organisations: OECD, World Trade Organization (WTO), IMF, and the World Bank (Harvey 2007, 3; Plant 2009, 5; Peck 2010, 1; Streeck 2011, 17).[3] This distrust of state-centred governance has led many economists and politicians to defend climate policies that are independent of political control (Buch-Hansen and Wigger 2011, 139).

As we saw earlier, the market cannot successfully internalise the externalities of climate change into prices without regulation. Hence, instead of suggesting free-market policies, different state-governed policies that imitate the mechanisms of the market have been implemented. There is a clear tendency today to subject many climate policies to market competition (Plant 2009; Peck 2010). The primary example is the cap-and-trade system.[4] In the European Union, the

Emissions Trading System is a cornerstone of the EU's efforts to meet its obligation under the Kyoto Protocol (IMF 2007). Another example is the idea of a *green
growth* policy framework, which aims to build a green economy where the main
drivers of growth are private investments in natural capital.[5] Yet other examples
are the REDD+ programme and the Green Climate Fund, which are the financing models of climate projects adopted in the Copenhagen Accord (Clapp and
Thistlethwaite 2012).[6]

These policies are considered to be the most efficient policy strategies given the
incentive critique that people are not expected to act according to moral, social,
and green values. Such economic incentives are believed to change the behaviour and self-interests of individual agents and private stakeholders. As discussed
earlier, the claim is that fulfilment of people's economic incentives will lead to
the highest aggregated outcome. The implication of this approach is that climate
politics should not be conducted and designed in democratic institutions. Instead,
climate politics should follow the logics of the markets as they are understood by
economic experts and economic reasoning (Scharpf 2009, 14; Buch-Hansen and
Wigger 2011, 139).

This critique may seem warranted if the challenge of climate change is assumed
to be one of motivation and economic incentives. I argued earlier that the framework of *the tragedy of the commons*, which understands climate change as a
challenge of economic motivation and incentives, is insufficient to comprehend
higher-level problems related to climate governance. In contrast, I also argued
that the framework of *the tragedy of the few*, which focuses on climate change
as a challenge of institutional design and social conditions, is more suitable.
Nonetheless, let us for the sake of argument consider more carefully the policy
understanding underlying the incentive critique. Is it plausible that these policies
provide better control over climate governance than policies that rely on democratic legislation, participation, and deliberation?

Critique of cognitive and moral capacities

As we saw earlier, Hayek provides a sophisticated theory of knowledge dissemination and economic functionality and assigns specific tasks to markets. This
scepticism regarding political planning is based on a general sceptical anthropology that questions what human beings are capable of doing, knowing, and planning. Accepting this scepticism implies a general precautionary approach to all
types of human activities and to all types of policies in particular.

The neoliberal critique of democratic climate governance cannot, however,
presume the sceptical anthropology in the Hayekian sense because the suggested
alternative policy solution presumes that a political system is capable of organis-
ing and controlling the market manipulations. One way to make sense of this argument is to comprehend the assumed sceptical anthropology not in the Hayekian
sense, but in the sense of Robert Nozick. Here, the underlying scepticism towards
people is not aimed at people's *cognitive* capacities, but their *moral* and *social*
capacities (Persson and Savulescu 2012, Chapter 2). This sceptical anthropology

is vindicated by studies in psychological and genetic dispositions. These studies indicate that people are ill-equipped to deal with the challenges of climate change due to humans' Stone Age mindset, which constrains the moral motivation of, for example, paying the costs of mitigating climate change for the benefit of future generations (Persson and Savulescu 2012, 1–2).

Based on this exposition, we can distinguish two types of critiques of democratic climate governance: a liberal *moral motivation* and a neoliberal *economic incentive* critique (Amable 2011; Scharpf 2006). Liberalism is the theory of knowledge deficits; it provides an argument for the necessity of the market distribution of goods, resources, and services. As we have seen earlier, the Hayekian liberal critique of democratic climate governance originates in the critique of political planning because of knowledge problems of complex societies and is therefore not a critique specific to democracies. The neoliberal critique takes the point of departure from a general distrust in political agents' moral and social capacities to process collective action problems efficiently. Thus, *the incentive critique* does not stress the knowledge deficit, but moral and social deficiencies, and with this in mind offers an argument for manipulating people's inherited egoistic motivations by competition.

Hence, the incentive critique of democratic climate governance recommends market-based manipulations of people's incentives. Note that this market ideology is in principle compatible with democratic governance. Indeed, the programme requires political institutions to implement the policies. Nonetheless, the policy design of market-based policies is not produced and constructed in democratic parliamentarian forums but frequently in economic expert committees (Scharpf 1996). Moreover, as we shall see next, the general distrust of citizens and politicians to do anything good for the climate provides the rationale to introduce different type of incentive manipulations.

In other words, the conceptualisation of the relevant set of climate policy options remains unclear. On the one hand, the neoliberal critique addresses the impossibility of democratic climate governance to govern climate change due to assumptions of lack of economic incentives. On the other hand, the critique presumes the possibility that the same group of politicians can govern market failures in relation to environmental harm and climate change by governing people's incentives in the market. But who manipulates the politicians' incentives? If politicians are capable and willing to regulate the market and introduce quasi-markets, why are they not capable of regulating climate change with other politically designed policies? And if they are, why are manipulations of the incentives at the market necessary in the first place?

Proactionary policies

In this section, I discuss the *moral motivation* critique of democratic control with climate governance. Proponents of the moral motivation critique put their faith in neither democratic politics nor the market to compensate for the moral deficiencies of humanity and the control deficiencies of contemporary democracies (Shearman and Smith 2007, 165; Persson and Savulescu 2012, 1, 122). They appreciate the *incentive critique* but expand the critique to exclude market-based solutions as well. The critique assumes that people do not have a moral nature that would make them capable of controlling the policy option that ought to be chosen with regard to climate change and suggests making changes in people's nature through biomedical modification.

These policy options are *proactionary* solutions which are the opposite of *precautionary* approaches. Precautionary policies aim to prevent the worst outcomes, whilst proactionary policymakers seek the promotion of the best available opportunities. Contrary to the worries of the precautionary camp, the proactionary approach is to encourage people to transcend current norms rather than adhere to them. Proactionary policy suggestions experiment with new technologies and new pharmacological prospects (Fuller 2012, 1–2).

Proactionary policies are not inherently incompatible with democratic climate governance. But the reason why proactionary policy initiatives are suggested as necessary is caused by the control deficiencies of democratic institutions. Thus, the underlying understanding of the problem, which provides the justification of proactionary policies, renders democratic climate policymaking redundant. For example, Persson and Savulescu suggested the controversial and radical *medical moral enhancement*, which is supposed to augment empathy and sympathetic concerns about the well-being of others and future beings (Persson and Savulescu 2012, 109). Two drugs are suggested as relevant for moral bioenhancement: the hormone and neuorotransmitter oxytocin and glucocorticoids. Studies of the drugs' influences on moral sentiments are as yet non-conclusive; some studies even show anti-group moral sentiments. Nonetheless, the authors stressed that we should not preclude biological moral enhancement as a possible way to reverse neglected climate action and to regain control with climate governance (Persson and Savulescu 2012, 133).[7]

Another type of bioengineering aims to change people's incentives for eating red meat. One big problem of climate change is the methane gas released from livestock farming. It is estimated to account for at least 37 percent of the world's greenhouse emissions. A reduction in livestock farming would have a positive environmental impact.[8] The idea is to induce a pharmacological meat intolerance – i.e., a vomiting substance that would decrease people's excessive desire for red meat. By inducing a vomiting substance into red meat, the idea goes, people will voluntarily avoid the environmentally unfriendly meat, causing a net benefit for human beings and nature.

Yet others defend a modification not of human moral or bio-capacities but of the earth's biosystem. Several geo-engineering strategies have been suggested. One much-discussed idea suggests a modification of solar radiation by injections of sulphur dioxide gas into the stratosphere (Lovelock 2009, Chapter 5; Gardiner 2010, Chapter 10; Hamilton 2010). Liao et al. argued that human engineering may be less risky than geo-engineering. Moreover, they emphasise, as do Persson and Savulescu, that equivalent pharmacological and biomedical forms of human enhancement and modification are already available. Based on this, they concluded that opponents of human engineering do not have a strong case.

Despite their differences, proponents of bio- and geo-engineering of human and environmental conditions share a radical willingness to engage in proactionary policies in order to save future generations and the earth from climate change by restoring the political control of climate politics. Yet I cannot help but ponder that whilst the urgency of climate change might allow for experimenting with human and environmental conditions, certain such experiments might possibly disregard the moral integrity of mankind and the earth and therefore also of democratic political institutions.

Several points need to be considered in order to evaluate such proposals on moral grounds. First, we should look into how such policies are justified. One way of justifying the proactionary policies is voluntarism. The policy of inducing meat intolerance is claimed to be a voluntary practice: Liao et al. argued that 'human engineering would be a voluntary activity [. . .] rather than a coerced, mandatory activity' (Liao et al. 2012, 211). Their argument for voluntarism, however, rests on thin ground: several important concerns are belittled and neglected. Liao et al. argued that human engineering is necessary because 'people often lack the motivation or willpower to give up eating red meat even if they wish they could' (Liao et al. 2012, 212). However, if individuals lack motivation and willpower, consumers cannot be expected to have the willpower to choose the modified meat instead of normal meat or no meat at all. Second, other proponents of proactionary policies do not even presume voluntarism. Persson and Savulescu argued that individuals may be subjected to the proactionary policies without their consent:

> [S]ome children should be subjected to moral bioenhancement [. . .]. This is because the capacity to influence development under way is likely to be greater than the capacity to alter established motivational dispositions and behaviour. There is no reason to assume that moral bioenhancement to which children are exposed *without their consent* would restrict their freedom and responsibility more than the traditional moral education to which they are also exposed without their consent.
>
> (Persson and Savulescu 2012, 113, emphasis added)

Three further points should be added. (1) First, this argument drifts towards the authoritarian argument, which will be discussed next. (2) Second, a point with regard to human nature should be added. If the nature of humanity is changed by pharmacological means, it is uncertain to what extent it will make sense to talk about voluntarism of human agents in the future.[9] If we change the nature of humanity fundamentally, we need to reconceptualise what human agency means. Moreover, policies that disregard the moral integrity of human nature, human rights, and the moral value of democratic participation are unlikely to respect voluntarism.[10]

Not all of the proactionary policy suggestions change people's nature. But even if the nature of human agency is not changed, other problems remain. For example, (3) how would these proactionary policies be implemented? *Who* should decide *when, how,* and *whom* to subject to these polices? Little is said about this. (4) Moreover, human- and geo-engineering are likely to be large-scale projects in need of enormous investments. Robock suggested in his account '20 Reasons Why Geoengineering Is a Bad Idea' that commercial and military interests are likely to control such projects (Robock 2008, 17). Hence, although the proponents of proactionary policies maintain the ideal of voluntarism, benevolence, and democracy, there are several reasons why proactionary policy programmes are ill-suited for maintaining voluntarism and strong democratic participation. Indeed, it is most likely that it will take authoritarian rule to enforce the suggested biological modifications of people's moral capacities and motivations and the geological modifications of the earth's climate and nature. Therefore, it also does not come as a surprise that the argument for proactionary policies drifts towards the argument for a benevolent dictator.

The authoritarian solution

Several of the proactionary scholars explicitly embrace different types of 'good authoritarianism' as a way to restore political control of climate politics (Persson and Savulescu 2012; Stehr 2013). Nonetheless, it is relevant to distinguish the proactionary policy suggestions, which aim to change people's motivations, from the authoritarian policy suggestions, which we now take a closer look at. The reason for this distinction is that the two aspects rely on two rather different and, as we shall see, not fully compatible sets of assumptions and therefore also on two different critiques of contemporary democratic climate governance. As we have seen, the point of departure of the proactionary critique presumes a universal sceptical anthropology: human nature makes individuals unlikely to do something good for the climate. If this sceptical anthropology is accepted, it is unlikely that the fitness argument for responsibility will be successful because political institutions will inherently be unable to control the relevant policy options necessary for curbing climate change.

The proponents of proactionary policies re-introduce a level of political control by suggesting different types of authoritarian decision making. They criticise the possibility of solving the ostensible motivation deficits by political and democratic means. Their concept of climate solution presumes a level of political authority that is in control of implementing proactionary policies – i.e., by designing specific technological interventions. If there are reasonable arguments for why authoritarian political bodies have better control over relevant climate options, there might be a serious test of the fitness for responsibility at the level of democratic climate governance. If democratic institutions have a substantively less capacity to engage in relevant climate politics, one might consider whether non-democratic institutions are fit to be held collectively responsible.

Let us now look into two of the arguments for authoritarian proactionary policies. The premise of both critical points is that there are two types of human beings. I take this premise to be inherently incompatible with democratic values because it assumes that some unfortunate people have a 'less morally motivated nature' and other fortunate people have a morally motivated nature (Persson and Savulescu 2012, 113). Assuming this, the critique of democratic governance does not rely on a universal (sceptical) anthropology which renders democracies unsuitable for climate governance, as we saw earlier in the *incentive* critique. The problem of democracy is not a universal sceptical anthropology, but that those in political power are not the wise and the benevolent ones (Shearman and Smith 2007). Hence, in order to restore the control of climate politics, it is not enough to replace democratic decision making with *market* or *voluntary* proactionary policy frameworks. In contrast, a *benevolent dictator* or *meritocratic elite* are considered as the only option. Democracies are thought to lack control of climate politics because they allow too many unwise and malevolent people to influence politics, which slows and derails the decision-making process.

Whilst neither the market-based nor the proactionary policy proposals are inherently compatible with democratic decision making, the authoritarian critique suggests *anti-democratic* solutions to climate governance. The benevolent dictator or elite is the answer for those who have lost patience with 'the seemingly interminable machinations of polycentric politics' (Dryzek and Stevenson 2011, 1865). Other scientists, who are also climate activists, proclaim that

> even the best democracies agree that when a major war approaches, democracy must be put on hold for the time being. I have a feeling that climate may be an issue as severe as a war. It may be necessary to put democracy on hold for a while.
>
> (Lovelock 2010)

The claim is that the democratic form of governance is incompatible with the governance of climate disasters, and a benevolent dictator or elite is thought to be more capable of managing and controlling the current state of affairs in an environmentally and climate-friendly way.

The eco-authoritarians of the 1970s had a similar argument. Some democratic rights and rules 'would have to be sacrificed to achieve sustainable future outcomes since authoritarian regimes are not required to pay as much attention to citizens' rights in order to establish effective policy in key areas' (Held and Hervey 2009, 5).[11] As Ophuls argued, 'Scarcity in general erodes the material basis for the relatively benign individualistic and democratic politics characteristic of the modern industrial era' (Ophuls 1977, 163). These anti-democratic arguments are not justified in voluntarism, as some of the proactionary arguments mentioned earlier were, but in the *benevolence* of the authoritarian regime. In other words, it is assumed that the authoritarian regime will rule in favour of the benefits of all and nature: one suggestion is that the authoritarian regime should be governed by a well-educated intellectual elite class with a special university degree (Shearman and Smith 2007, Chapter 9).

The factual aspects of the critique

Thus far, examples of the critique of democratic control of climate governance have been presented, and the *incentive*, *motivation*, and *authoritarian* critiques have been distinguished. It should be noted that the critiques are not proposed as a normative critique of democratic institutions and values. They are justified instead as second-best options, given the fact that democratic institutions lack control of satisfactory climate policies (Bevir 2010, 575). The necessity of critique is defended in the factual circumstances of politics: the current failures of democratic climate governance do need to be addressed. However, the factual component of these critical points has not been adequately demarcated. In this section, I discuss two aspects of the various critiques. I argue that we might agree to the factual components of the critique of democratic climate governance without accepting the necessity of initiating neoliberal, proactionary, and authoritarian policy experiments.

The critique of democratic climate governance takes its point of departure from the factual claim that the current 'Western societies are not democracies as such but plutocracies, societies ruled by the wealthy' (Shearman and Smith 2007, 91), and that 'traditional liberalism has been too permissive as regards letting citizens of affluent societies adopt ways of living that waste the resources of the planet' (Persson and Savulescu 2012, 122). There is a multitude of forces that are acting to corrupt them and prevent them from truly representing the voice of the people. Various powerful elite groups rule the modern liberal democracy, being based in finance, media, business, and the military, and they have their own agenda that does not advance the interests of the common good among them. (Shearman and Smith 2007, 89)

Further facts support this critique. Indeed, several empirical studies of the current status of democratic quality support the elitist component of the authoritarian critique of democratic climate governance (Streeck 2011, 24). Recent studies show that the quality and transparency of contemporary democracies is deteriorating, the number of full democracies is decreasing, and, moreover, the strength of the full democracies is deteriorating (Scavenius 2012). In 2011, France, Italy, Greece, and Slovenia dropped from the category of full democracies to the category of flawed democracies, a category that includes South Africa, Botswana, Columbia, and Thailand (EIU 2011, 5–10).[12]

Yet other empirical studies on the correlation between democracies and environmental outcomes are non-conclusive. Democratic societies do not unequivocally have a better climate record than non-democratic societies. Some results show that there is a strong correlation between democracy and high environmental quality. Among the 40 highest carbon emitters, the countries that have the best records are all democracies (EIA 2006; Held and Hervey 2009, 6–7).

Other results are less conclusive. Manus Midlarsky, for example, finds that democracies have good records in land area protection but not in deforestation, CO_2 emissions, and soil erosion (Midlarsky 1998).

These aspects of the critique of democratic climate governance fit well with the earlier discussed *tragedy of the few* dilemmas in which the political problems of soil degradation, air pollution, and carbon pollution should not be found in the free access to the commons (cf. *the tragedy of the commons*). In contrast, contamination and overconsumption are caused by unequal control over resource subtraction and pollution.

It is worth noting here that the often harsh and strongly-worded critique of democratic governance is supplemented with few lofty ideas for solving the problem of motivation or incentives (Persson and Savulescu 2012, 1; Shearman and Smith 2007, 165). No comprehensive sustainable policy design is suggested. Persson and Savulescu (2012) and Liao et al. (2012) suggest pharmacological treatments that do not yet exist and admit that they are unlikely to become available in the near future. Shearman and Smith (2007, 152) provide an underdeveloped idea of a new education of the meritocratic elite.

Even if they did expand the concrete policy suggestions, these critics fail to provide a comprehensive political theory that can compete with the political theory of democratic institutional design and decision-making processes. They underestimate in particular the importance of including ideas about how the institutions are thought to be maintained. Common in the critiques is that they presume the existence of well-functioning national and global institutions. Note that the market-based policy proposals presume the existence of a well-functioning and effective global market and the proactionary critique presumes the existence of a political system that has the willingness and capacity to conduct trans-human and trans-geological experiments in a voluntary and benevolent way. And the authoritarian regime suggestion presumes the existence of a well-educated meritocratic elite class that has the will, capacity, and power to administer and rule society by environmentally friendly means. Nothing is said about how to maintain the social, political, and institutional conditions of the suggested climate policies.

Put differently, these three very different critiques converge in their neglect of the importance of maintaining the political institutions that make it unlikely that they would provide a stronger control with climate governance compared to democratic institutions that include participation and deliberation. One reason for this is that democratic institutions satisfy the second fitness-condition of how institutionalised group agents have access to the required knowledge and evidence. Thus, the democratic theory is not merely a theory of decision-making processes, nor is it a theory of one technological fix (Streeck 2011). It is also an epistemic theory of what collective groups of people are capable of doing if they coordinate their acts and behaviours collectively and institutionally, as discussed earlier, and a political theory with dynamic and flexible instruments to sustain and enhance its institutional design and capacity.

Hence, if it is true that democratic institutions are corrupted by corporate and elitist interests, and if *the tragedy of the few* resembles the challenges of climate governance, the critique of current democratic control with climate governance must be accepted – not for normative but for factual reasons. However, it remains to be discussed what conclusions we might draw from this. One alternative option to the policy proposals discussed in this chapter is to restore democratic control of climate governance with democratic means to distribute access to and the benefits of natural resources through decision processes that benefit from democratic institutional capacities.

Democratic reforms

I note that the recent history of democratic governance does not univocally stipulate reasons for the sceptical anthropology underlying the critique of democratic governance (Le Grand 2006, 4). There are many examples of successful national and international cooperation on collective action problems of security, environmental matters, and public infrastructure and services. This suggests that there is no simple correlation between one democratic institutional design and whether climate policies are successful. Thus, it is impossible to critique democratic decision making and climate governance *en bloc* because, as recent empirical studies have shown, people's way of behaving varies according to institutional designs and behavioural spaces (Frey and Jegen 2000; Ostrom 2005; Le Grand 2006, Chapter 3).

One conclusion in particular provided by Julian Le Grand suggests that people sometimes act according to economic motivation and sometimes according to moral motivation. He concluded that financial rewards for something people are morally motivated to do 'erode the magnitude of the sacrifice that he or she is making, and thereby partly erode the motivation to act' (Le Grand 2006, 67). Le Grand did not argue that motivational structure should not be part of a policy package: he argued that policymakers should be aware of when one policy is designed to align economic or moral motivation. Only then it is possible to identify when a particular economic or moral incentive design should be implemented or not.

Specifically, it is important to allow for a political system capable of channelling people's moral motivations with regard to climate change through democratic decision-making processes. Thus, when we point to the challenges related to *the tragedy of the few*, the critique of democratic decision making looks fundamentally different. Of foremost importance is the fact that the current problems that democratic climate governance face may be caused by climate politics that focus primarily on the agent's economic incentives, which potentially erode the relevant agents to act at all. So conceived, the only way out of a climate disaster is not incentive manipulation at the market or through pharmacological means. Instead, it may be to enhance and improve the democratic institutional capacity through which people's moral inclinations can flow freely.

Thus, before discarding democratic institutions as unfit for climate governance, democratic reforms of the current democratic environment should be considered as possible strategies to restore the political control of climate policies. Hence, although we accept the factual aspects of the neoliberal, proactionary, and authoritarian critiques that many contemporary democracies have failed to govern the climate satisfactorily and no longer are *real* and *full* democracies, an alternative inference can be suggested which defends democratic reforms of the political elitism:

> Democratic critique: *Lack of democratic reforms challenges climate politics*. Democratic climate politics has become too elitist. Elitism undermines the institutional, deliberative, and participatory advantages of democratic decision making.

The argument, then, would state that rather than replacing democracy with a more authoritarian governmental authority, democratic reforms should be considered as a way of fortifying and invigorating the democratic quality of flawed democracies. This would arguably also enhance the fitness of contemporary democracies to provide climate governance. The argument goes like this:

1   We are facing a climate change with catastrophic impacts;
2   *Democratic* institutions have become too elitist, rendering them unable to counteract climate change politically; and
3   In order to curb climate change and avoid a catastrophic climate disaster, it requires democratic institutions that *can* solve the problems related to *the tragedy of the few*.

According to this argument, political institutions do not become more fit to govern the climate by centralising the political decision-making process in elitist structures or by trespassing the moral integrity of humanity, as suggested by proponents of the critiques of democratic climate governance. In contrast, political institutions can only manage the problems related to environmental harm and climate change if they are well-functioning, democratic, and capable of benefiting from democratic deliberation and participation by various agents.

To recall a point from the previous or discussion: no single agent can have accurate knowledge about what is the best climate politics and arguably no control over relevant climate policies. If we embrace this observation, an authoritarian leader can possess neither the necessary knowledge nor the necessary control. Thus, it is reasonable to conclude that democratic institutions are the fittest group agents to be held morally responsible for climate change. Nonetheless, the critiques of democratic climate governance teach us something valuable about the factual conditions and constraints for conducting democratic climate governance, and arguably, about the normative discussion of whether we can blame the democratic group agent for doing too little with regard to being capable of controlling the relevant policies. I turn to this question of blame and excuse in Part 3.

Conclusion

I have provided a detailed analysis of the control fitness-condition of moral responsibility with regard to the institutionalised group agents of contemporary democracies. I have also discussed critiques of democratic climate change. Two conclusions of this discussion should be distinguished: one is theoretical, the other political. The theoretical conclusion is that institutionalised group agents of democracies, if designed correctly, outmatch those of authoritarian decision processes. Such group agents in democracies can reap the potential benefits of free research, media, freedom of expression, dynamic political processes, and separation and diffusion of power. We can conclude on this basis that political-democratic institutions are institutionalised group agents that are fit to be held collectively responsible for grand challenges, such as climate change, because they meet the established three conditions: (1) access to knowledge, (2) morally significant policy choices, and (3) control over relevant policy choices.

In contrast, the conclusion looks different from the political point of view. One primary conclusion is that it is unclear whether the control premise is satisfied in contemporary democratic societies. One reason for this is that democratic climate governance has become too elitist, which negates the benefits of democratic decision making. Thus, if current democracies do not have control over relevant policy options because the potential epistemic gains and innovative initiatives of democracies are not utilised, the third fitness-condition for collective responsibility is not fulfilled. The failures of democratic climate governance reflect a deeper challenge of the institutional capacity of current political institutions.

So conceived, failure of climate governance is not merely a question of motivation and incentives but of deteriorating institutional capacity, which renders effective democratic climate governance impossible. Institutionalised group agents with low political capacity are less fit to be held responsible for rectifying environmental harm and climate change. Nonetheless, if democratic societies are currently incapable of controlling comprehensive climate change, the only plausible alternative is not a dismissal of democracy but a democratic reform of the current democratic institutions and arguably of their capacity to do climate governance.

This chapter completes the investigations of the fitness-conditions of rational individual agents and institutionalised group agents. In Part 3, I examine how the fitness-conditions feed into the consideration of who, if anyone or any entity, ought to be held morally responsible for climate change.

Notes

1 Two understandings of neoliberalism can be distinguished: one stresses the continuity between classic liberalism and neoliberalism and the other stresses the discontinuity between the two. For more, see Thorsen (2010) and Amable (2011).
2 The current political paradigm of liberalism is frequently attributed to the mid-century economic philosophers. Hayek, together with Ludwig von Mises, James Buchanan, Milton Friedman, and Robert Nozick had a tremendous impact on the policy paradigms of the 1980s, 1990s, and 2000s (Harvey 2007; Buch-Hansen and Wigger 2011).
3 The critical approach to state-centred politics grew out of the dysfunctions and inefficiencies of the excessive fascist or socialist states of the 20th century. For both economic and normative reasons, it can be argued that it was justified to limit the scope of the state and to turn over tasks, functions, and responsibility to the market and to civil society (Fukuyama 2004, 119).
4 Examples of international trading schemes are the Kyoto Protocol and the EU Emissions Trading System (EU ETS). Examples of national trading schemes can be found in the US, New Zealand, and Australia (IMF 2007).
5 The idea of green growth was first promoted by South Korea and the *United Nations Economic and Social Commission for Asia and the Pacific*. In 2012, the leading economic international organisations and institutions, OECD, UNEP, World Bank, and Global Green Growth Institute (GGGI), signed a Memorandum of Understanding to promote economic solutions to sustainability and climate change (Green Growth Knowledge Platform 2012). For a thorough discussion, see Jackson (2009).
6 The Copenhagen Accord was the outcome of the 2009 COP15. The REDD+ programme is a policy mechanism that aims to provide financial incentives for developing countries to Reduce Emissions from Deforestation and Forest Degradation, 'plus' sustainably manage forests and conserve and enhance forest carbon stocks (UNFCCC

2010). The Green Climate Fund was the outcome of the Cancun 2010 COP15 and provides a funding mechanism by which public and private investments can give grants or loans to projects in developing countries (UNFCCC 2010; Bird et al. 2011).

7 The pharmacological treatment is thought to be especially necessary for 'aggressive males', under the assumption that women have a greater biological capacity for empathy than men (Persson and Savulescu 2012, 111). One reason for this is that the hormone oxytocin is naturally produced during birth and breastfeeding, and mediates maternal care and other pro-social attitudes (Persson and Savulescu 2012, 118). If this is true, one plausible inference is not to enhance the anti-social attitudes of men but to hand over the political power to breast-feeding women. This argument resembles the gist of *ecofeminism*. For a classic account, see Warren (1987; 1990).

8 Liao et al. (2012) suggested, 'While reducing the consumption of red meat can be achieved through social, cultural means, people often lack the motivation or willpower to give up eating red meat even if they wish they could. Human engineering could help here. Eating something that makes us feel nauseous can trigger long-lasting food aversion. While eating red meat with added emetic (a substance that induces vomiting) could be used as an aversion conditioning, anyone not strongly committed to giving up red meat is unlikely to be attracted to this option. A more realistic option might be to induce mild intolerance (akin, e.g., to milk intolerance) to these kinds of meat' (Liao et al. 2012, 212).

9 Persson and Savulescu argued that moral bioenhancement does not make individuals unfree (Persson and Savulescu 2012, 112–115). For a critical discussion of this argument, see Harris (2011).

10 Note that the institutionalised group agent of these proactionary policies would not satisfy the first condition of responsibility. If the fundamental norms for moral universalism and human agency are eliminated, no normatively significance of choice of climate governance can be stipulated. Since the purpose in this part is to examine whether democratic institutions are fit to be held morally responsible for normatively significant choices (assuming moral universalism and the moral integrity of human agency), I set these considerations aside.

11 See also Hardin (1968) and Ophuls (1977).

12 The reasons for downgrading France to a flawed democracy are (1) deterioration of media freedom, (2) extremely low public confidence in political parties and government, (3) declining engagement in politics, (4) low degree of popular support for democracy; (4) a widening gap between the people and political elites, (5) violent rioting as a symptom of the country's political malaise, (6) power concentration around the president, and (7) increased anti-Muslim sentiments. The reason for downgrading Italy to a flawed democracy is the country's politised media. Greece has been downgraded because of low scores on government functioning and political culture. Furthermore, corruption has increased and government transparency and accountability are low (EIU 2011, 16–17).

References

Amable, B., 2011. Morals and politics in the ideology of neo-liberalism. *Socio-Economic Review*, 9, 3–30.

Bevir, M., 2010. *Democratic governance*. Princeton, NJ: Princeton University Press.

Bird, N., Brown, J. and Schalatek, L., 2011. *Design challenges for the Green Climate Fund, 18 January 2011*. Heinrich Böll Stiftung The Green Political Foundation. Available from: www.boell.de/en/ecology/ecological-economics-design-challenges-for-the-green-climate-fund-10976.html [Accessed 13 March 2016].

Buch-Hansen, H. and Wigger, A., 2011. *The politics of European competition regulation: a critical political economy perspective*. New York, NY: Routledge.

Clapp, J. and Thistlethwaite, J., 2012. Private voluntary programs in environmental governance: climate change and the financial sector. *In:* K. Ronit, ed. *Business and climate policy: the potential and pitfalls of private voluntary programs*. New York, NY: United Nations University Press, 61 100.

Dryzek, J. S. and Stevenson, H., 2011. Global democracy and earth system governance. *Ecological Economics*, 70 (11), 1865–1874.

Economist Intelligence Unit (EIU), 2011. *The Democracy Index 2011: democracy under stress*. Available from: www.sida.se/globalassets/global/about-sida/sa-arbetar-vi/eiu_democracy_index_dec2011.pdf [Accessed 3 April 2019].

EIA, 2006. *International energy annual 2006*. Washington, DC: United States Energy Information Administration.

Frey, B. S. and Jegen, R., 2000. *Motivation crowding theory: a survey of empirical evidence*. Zurich IEER Working Paper No. 26; CESifo Working Paper Series No. 245. Zurich: University of Zurich.

Fukuyama, F., 2004. *State-building: governance and world order in the 21st century*. Ithaca, NY: Cornell University Press.

Fuller, S., 2012. *The future of ideological conflict*. Project Syndicate, 7 May, 1–2.

Gardiner, S. M., 2010. Ethics and global climate change. *In:* S. M. Gardiner, S. Caney, D. Jamieson, and H. Shue, eds. *Climate ethics: essential readings*. Oxford: Oxford University Press, 3–38.

Green Growth Knowledge Platform, 2012. *Global organizations to expand cooperation on green growth for development*. News Release, 11 January, Mexico City. Available from: www.greengrowthknowledge.org/SiteCollectionDocuments/ Mexico%20City%20Confer ence%20Papers%20and%20Presentations/GGKP%20MoU%20and%20Conference%20 Press%20Release%20(English)-%20FINAL.pdf [Accessed 5 December 2015].

Hamilton, C., 2010. *Requiem for a species*. London: Earthscan.

Hardin, G., 1968. The tragedy of the commons. *Science*, 162 (3859), 1243–1248.

Harris, J., 2011. Moral enhancement and freedom. *Bioethics*, 25 (2), 102–111.

Harvey, D., 2007. *A brief history of neoliberalism*. Oxford: Oxford University Press.

Held, D. and Hervey, A. Fane, 2009. *Democracy, climate change and global governance democratic agency and the policy menu ahead*. Policy Network Paper, November. London: Policy Network.

IMF (International Monetary Fund), 2007. *World economic outlook: globalization and inequality*. Available from: www.imf.org/external/pubs/ft/weo/2007/02/ pdf/text.pdf [Accessed 23 January 2017].

IPCC (Intergovernmental Panel on Climate Change), 2013. *Summary for policymakers*. Twelfth Session of Working Group I Approved Summary for Policymakers. Available from: www.climatechange2013.org/images/uploads/WGIAR5-SPM _Approved-27Sep2013.pdf [Accessed 20 January 2017].

Jackson, T., 2009. *Prosperity without growth: economics for a finite planet*. London: Earthscan.

Le Grand, J., 2006. *Motivation, agency, and public policy: of knights and knaves, pawns and queens*. New York, NY: Oxford University Press.

Liao, S. M., Sandberg, A., and Roache, R., 2012. Human engineering and climate change. *Ethics, Policy and Environment*, 15 (2), 206–221.

Lovelock, J., 2009. *The vanishing face of Gaia: a final warning*. New York, NY: Basic Books.

Lovelock, J., 2010. James Lovelock on the value of sceptics and why Copenhagen was doomed. Interviewed by Leo Hickman. *The Guardian*, Monday, 29 March. Available from: www.theguardian.com/environment/blog/2010/ mar/29/james-lovelock [Accessed 23 September 2013].

Mathews, F., ed., 1991. *Ecology and democracy*. London: Frank Cass.

Midlarsky, M., 1998. Democracy and the environment: an empirical assessment. *Journal of Peace Research*, 35 (3), 341–361.

Ophuls, W., 1977. *Ecology and the politics of scarcity*. San Francisco, CA: Freeman.

Ostrom, E., 2005. *Understanding institutional diversity*. Princeton, NJ: Princeton University Press.

Peck, J., 2010. *Constructions of neoliberal reason*. New York, NY: Oxford University Press.

Persson, I. and Savulescu, J., 2012. *Unfit for the future: the need for moral enhancement*. Oxford: Oxford University Press.

Plant, R., 2009. *The neoliberal state*. Oxford: Oxford University Press.

Robock, A., 2008. 20 reasons why geoengineering may be a bad idea. *Bulletin of the Atomic Scientists*, 64 (2), 14–18, 59.

Saad-Filho, A. and Johnston, D., 2005. Introduction. *In:* A. Saad-Filho and D. Johnston, eds. *Neoliberalism – A critical reader*. London: Pluto Press, 1–6.

Scavenius, T., 2012. Transnationalism vs. nationalism: the case of free movement. *Global Justice*, 5, 82–93.

Scharpf, F. W., 1996. *Economic integration, democracy and the welfare state*. MPIfG Working Paper 96 (2).

Scharpf, F. W., 2006. The joint-decision trap revisited. *Journal of Common Market Studies*, 44 (4), 845–864.

Scharpf, F. W., 2009. Legitimacy in the multilevel European polity. *European Political Science Review*, 1 (2), 173–204.

Shearman, D. and Smith, J. W., 2007. *The climate change challenge and the failure of democracy*. Westport, CT: Praeger.

Stehr, N., 2013. An inconvenient democracy: knowledge and climate change. *Society*, 50, 55–60.

Streeck, W., 2011. The crises of democratic capitalism. *New Left Review*, 71, 5–29.

Thorsen, D. E., 2010. The neoliberal challenge: what is neoliberalism. *Contemporary Readings in Law and Social Justice*, 2 (2), 188–214.

UNFCCC, 2010. *Outcome of the work of the Ad Hoc Working Group on long-term cooperative action under the Convention*. Advance unedited version, Draft decision-/CP.16. Available from: http://unfccc.int/files/meetings /cop_15/application/ pdf/cop15_lca_ auv.pdf [Accessed 14 May 2015].

Warren, K., 1987. Feminism and ecology: making connections. *Environmental Ethics*, 9 (1), 3–20.

Warren, K., 1990. The power and the promise of ecological feminism. *Environmental Ethics*, 12 (2), 125–146.

Part 3

Moral responsibility for climate change

7 Collective responsibility

In this Part 3, I apply my fact-sensitive account of ought-assignment to agents, focusing particularly on the two normative criteria. As we saw earlier, the fact-sensitive account of moral responsibility contains two core components; one *methodological* concern with regard to the can- and fitness-conditions of the relevant moral agents and one *normative* concern in regard to the (1) moral blame and (2) moral justification of assigning ought-judgements to particular groups of agents. To reiterate, the fact-sensitive account of moral responsibility defines what *A ought to do* by the following conditions where each point is necessary and jointly sufficient:

OIC#1

A ought to do X

a   If and only if there is a moral wrongdoing;
b   If and only if A is a moral agent;
c   If and only if A can do X;
d   If and only if A is not morally excused; and
e   If and only if there is a robust justification of why A should be held morally responsible.

In the introduction, I outlined why climate change is a case of moral wrongdoing, and Part 2 has been occupied with whether moral agents can do what ought to be done. I have examined whether the methodologies of rational-individualism and non-reductionist institutionalism provide a framework in which agents are fit to be held morally responsible for climate change and concluded that the framework of rational-individualism does not satisfy conditions two and three with regard to climate change. Thus, it is redundant to consider whether the rational agents are morally blameworthy. I argued instead that the methodology of non-reductionist institutionalism provides a framework in which institutionalised group agents of democracies are fit to be held morally responsibility for climate change.

In Part 3, I shift the focus from whether group agents at the collective level of agency *can* be held responsible for whether they *should* be held responsible.

As mentioned earlier, there is a renewed interest in how to develop a theory that allows for responsibility at the collective level of agency (Vanderheiden 2011; S. Miller 2011). Yet there remains disagreement about what the relevant properties of different conceptualisations of responsibility are.[1] One important disagreement concerns what methodological framework the collective level of agency is approached. If an individualist methodology is presumed, institutions and group agents are perceived as nothing but the aggregate of individual interactions (S. Miller 2010, 121). In contrast, some collectivists argue that collective responsibility refers to emergent collective entities where, for example, a national culture is held collectively responsible and upholds what some have called a 'metaphysical guilt' (Vanderheiden 2011, 214).[2]

When we embrace the non-reductionist institutionalism that I elaborated on in Chapter 3, the collective level of agency qualifies as morally autonomous without being emergent collective entities. The presumptions of the account of collective responsibility, which I defend in this chapter, are thus the earlier established properties of institutionalised group agents, that is, that they are multi-level complex institutions with disconnected input and output sides. In Chapter 3, I invoked the example of the brain and its neurons as an analogy to understand the relationship between the collective level of agency of an institutionalised group of agents and its constituencies. In the example of democratic institutions, the constituencies are every individual agent that plays a role that is constitutive of democratic institutions. It is not collective entities that are assigned responsibility in this methodological framework: individual agents are held responsible in their capacity of upholding roles constitutive for institutionalised group agents.

As we saw earlier, the social institutions of the marketplace do not uphold a societal role for the rational agent in which they are fit to be held morally responsible for climate change. In contrast, I have argued that democratic institutionalised group agents are organised in a way that allows for moral agency. Engaging democratic institutions as an institutionalised group agent allows for several roles for which they are fit to be held morally responsible. In this context, I am particularly interested in the role of citizens and public officeholders, such as public servants, politicians, and ministers. I investigate whether we hold *democratic citizens* and *occupants of institutional roles* in modern democracies morally blameworthy for not providing the conditions where climate change can be mitigated and circumvented.

Joint and collective responsibility

In this chapter, I discuss two of the key concepts that underpin moral responsibility at the collective level of agency: *joint* and *collective responsibility*. In order to assign group agents responsibility for climate change in their capacity of occupying different societal and institutionalised roles, we need an account of how responsibility is distributed within the group. At this point, responsibility at the collective level can be defined in two ways. Collective responsibility can be applied in cases where each individual in the group is *collectively* responsible for an outcome or where none of the individuals in the group are responsible for an outcome.

To differentiate the two, I use different concepts. In cases where groups of individuals are jointly responsible for an outcome, I talk about *joint* responsibility. Each individual is responsible for his or her share of the outcome and it is possible to track each individual contribution. This means that there are two ways of assigning responsibility to institutionalised group agents, one in which each individual agent constituting the group agent is *jointly* responsible and one in which the *institutional role occupant* is *collectively* responsible. When we hold occupants of institutional roles morally responsible, they are not held individually responsible; instead, they are held collectively responsible as representatives of the institutionalised group agents (S. Miller 2010, 52–54).

Joint responsibility is a kind of *outcome* responsibility that is closely related to people's intentions and acts. Outcome responsibility assumes that agents can be held accountable for the outcome of their intended acts. Thus, outcome responsibility can be conceptualised both as individual responsibility, which holds one agent individually responsible for the CO_2 emissions she emits when driving her SUV, and joint responsibility, which holds a group of agents jointly responsible for the fact that a country's CO_2 level is higher compared to other similar countries. Many call individual and joint outcome responsibility as *backward-looking* responsibility, where attention is drawn to who has responsibility – that is, who has done what in the past (van de Poel 2011, 37).

Backward-looking responsibility can be contrasted with *forward-looking* responsibility, which indicates an interest in those who ought to do what ought to be done in the future. My conceptualisation of collective responsibility is a forward-looking responsibility, which assigns moral responsibility to those agents that should remedy the negative effects which have been produced in the past (D. Miller 2007, 83–84; Vanderheiden 2011). When we employ the notion of collective responsibility, we are less interested in tracking single packages of CO_2 emissions to single agents – i.e., in order to hold agents individually and jointly responsible. The concern instead is to assign moral responsibility to those agents who ought to do something about it.

When we address the forward-looking question of collective responsibility, it is unfair to take citizens of contemporary democracies to be the *primary* responsible agents in modern complex modern societies (Lichtenberg 2010, 76). I argue that it is fairer to assign moral responsibility to those who are collectively responsible for not choosing climate politics with high positive effects – namely, the occupants of the institutional positions of modern democracies. This is because whilst we have reasons to excuse the individual agents for some of their moral deficits, these reasons do not apply to what occupants of institutionalised roles are collectively responsible for.

In the case of climate change, it might be possible to assign some joint responsibility to the group of democratic citizens despite the fact that, as we shall see next, their climate-friendly actions have low positive impact on the climate because of the way the technological, industrial, and economic infrastructure of most modern democracies are constructed. Thus, if the goal is not merely to point fingers at particular agents but to suggest effective solutions to how climate politics can be organised in a more fruitful way, it is more productive to draw attention to that group of agents which has the capacity to bring forward real policy changes that have high positive effect on the climate.

By so arguing, the assignment of collective responsibility to particular groups depends on their capacity to be fit to be held morally responsible for climate change by deploying the fact-sensitive theory of moral responsibility elaborated earlier. I defined three fitness-conditions earlier – namely, whether the agents face a normatively significant choice and have access to sufficient knowledge and control over the choice situation. Whereas the notion of capacity and effectiveness importantly feeds into the normative discussion of moral responsibility for climate change, it is an insufficient account of who should be assigned responsibility. The subsequent two chapters provide the normative justification of why it is fair to assign ought-judgements to the groups' agents that have the highest capacity and thus potentially the highest level of impact.

The blame criteria

In this section, I elaborate on the first normative condition for the assignment of moral responsibility for climate change to institutionalised group agents – namely, the notion of moral blame and excuse. We can distinguish standard consequentialist moral justifications of acts and policies from non-consequentialist ones. Consequentialism is a normative theory that claims that the rightness or wrongness of one act is determined by the goodness or badness of its consequences. In contrast, intentionalism and other types of non-consequentialism are normative theories that deny that the measurement of rightness and goodness can be determined solely by the goodness or badness of the consequences one act may have (Kamm 2007, 11).

In the context of climate change, the pertinent question is whether we should blame groups of agents for the intentions they have whilst acting in a climate (un) friendly manner or whether we should blame the agents by evaluating the total outcome of their climate (un)friendly actions. Consider one example. National governments and the European Commission have allowed farmers to use several types of fertilisers on the crops. Let us say that the intention of the politics was, among other things, to enhance production and food security. In many cases, however, the fertilisers were found to cause human health problems and environmental damage. When one calculates the goodness (food security) and the badness (health problems and environmental damage) of this example, the result may be that the total badness is higher than the goodness of the fertiliser policy. This case illustrates the fact that many acts and policies that may have harmful and lethal consequences are put forward with good intentions.

When we invoke a non-consequentialist account of the fertiliser policy, the occupants of the relevant institutional roles in national governments and the European Commission should not be held morally responsible for the negative effect the fertiliser policy has on human health and the environment. In contrast, if we invoke a consequentialist account of the fertiliser policy, the relevant occupants of institutional roles should be held collectively responsible for the total negative outcome of the policy. In other words, they should be morally blamed for not changing policy or rectifying the negative effects of the fertiliser policy.

This normative approach to collective responsibility has been neglected in the literature to a large extent. One reason for this is that consequentialism has been conflated with utilitarianism, which to many provides an unsuitable moral theory of agency. Although any kind of utilitarianism is by definition consequentialist, utilitarianism is only '*one sort* of consequentialism which is especially concerned with happiness' (Geuss 2008, 79 emphasis in original). In contrast, consequentialism is not merely a theory about maximising utility or happiness but is 'the doctrine that the moral value of any action always lies in its consequences, and that it is by reference to their consequences that actions, and indeed such things as institutions, laws and practices, are to be justified if they can be justified at all' (Geuss 2008, 79).

Another more important reason for the neglect of consequentialist accounts of collective responsibility is that consequentialist accounts of morality have been dismissed at the level of individual agency (which I also defend in the next chapter). However, since many scholars, particularly proponents of rational-individualistic methodologies, do not distinguish between agency at the individual and collective level, the relevance of examining remedial collective responsibility in consequentialist terms has been disregarded. However, by embracing a methodological difference between these two levels of agency, consequentialism can be considered a fruitful strategy for evaluating social and political morality without neglecting the moral significance of taking individual agent's intentions and desires into account (Goodin 1995, Chapter 1; Geuss 2008, 135).

Consider the example of the fertiliser programme again. If we accept a consequentialist account of fertiliser politics, it provides an important tool that allows for the assignment of responsibility to the relevant group agents for the unintended consequences of the political decisions that were implemented in good faith. Evaluating the goodness of a public policy by its consequences allows for the assignment of collective responsibility to group agents that did not intend the bad consequences to happen. Likewise, with regard to climate change, the moral assignment of ought-judgements to particular agents cannot be fully captured by a non-consequentialist account since climate change constitutes a bad outcome that nobody intended. Thus, if we do not allow for a consequentialist account of whom it may be fair to assign responsibility to, nobody may be remedially responsible.

The point I want to pursue is that many reject a consequentialist approach to moral responsibility by pointing out that utilitarianism is an unsuitable approach to the moral evaluation of individual agent's acts and choices (Goodin 1995, Chapter 1). However, if we reject that consequentialism should be conflated with utilitarianism and moreover, that the individual level of agency does not exhaust whom it may be fair to assign moral responsibility to, a different conclusion can be drawn. This conclusion says that it is fair to take a consequentialist approach to the policies, choices, and acts of institutionalised group agents. Appreciating this, I believe it is appropriate to apply the different accounts to different types of agents. I suggest that it is fair to take an intentionalist account of individual and joint responsibility, and that it is fair to take a consequentialist account of collective responsibility. This is possible when we accept the earlier elaborated non-reductionist methodology, which allows for compatibility of individual and collective levels of moral agency.

At the collective and institutional levels, agency is not conducted by sullied agents but by a multi-level and complex group agent in which the input- and output sides are non-reducible to each other. By so arguing, it is possible to justify an intentionalist account of individual agent's intentions and motives to act on the input side and at the same time justify a consequentialist account of the aggregated outcome of these acts at on the output side. By invoking this account of moral agency, it is possible to apply the different conceptualisations of responsibility to particular agents depending on whether they are fit to be held *individually*, *jointly*, or *collectively* responsible for, in this case, climate change. Recall that individual and joint responsibility is defined as a backward-looking outcome responsibility, and collective responsibility is defined as a forward-looking remedial responsibility.

In order to provide a plausible account of which concepts of responsibility apply to particular agents, let us consider the fertiliser example again as an example of a choice between allowing lethal fertiliser to improve the output of the harvest and banning these lethal fertilisers because the health problems and environmental damage they cause. The question is to whom it may be relevant to assign moral responsibility for the total outcome of the fertiliser politics. If we maintain the assumption that nobody intended the health problems and environmental damage to happen, it seems inappropriate to invoke the principles of individual or joint responsibility of individual agents, such as the individual citizens and occupants of institutional positions. In contrast, I take it to be fair to assign collective remedial responsibility to occupants of relevant institutional positions for the total consequences of their politics and moreover, to blame them if they do not try to rectify these negative effects. Thus, moral agents should be held responsible for what they can do if we do not have reason to morally excuse them.

Democratic climate governance

Until now, I have argued that according to the fact-sensitive account of ought-assignments, those agents that *can* are not morally excused and *should* be held morally responsible. In this section, I outline the second normative criterion about to whom it is fair to assign responsibility. In Chapter 9, I develop this criterion further. The notion of fairness depends on the account of what the relevant moral agents are capable of doing. We should be worried when the third fitness-condition of control over policy options was not satisfied for contemporary political democracies, as argued in Part 2. If the quality and institutional capacity of contemporary democracies are deteriorating, it might be pointless to request ambitious climate action and arguably to hold the institutional role occupant responsible for not doing enough.

If the occupants of institutional positions *cannot* on behalf of the institution-alised group agent do anything that has high climate impact, they should not be held responsible. Some may argue that it does not matter because other, for example, authoritarian, regimes are capable of doing what is necessary, and one might suggest that they ought to be allowed to do so. Notwithstanding my approach the political problems of climate change, authoritarian and elitist regimes do not have the capacity to manage the normative and factual challenges related to environmental harm and climate change.

When we appreciate that one of the major problems of climate change is *the tragedy of the few*, we can appreciate that it is *only* democratic political institutions that have the potential to do what ought to be done. If their fitness to be held responsible is deteriorating, there may be no other agents that are suitable for the task of curbing the negative effects of climate change.

Thus, the first step towards better climate governance may be empowerment (1) of people's independence and competence and (2) of the democratic institutional design to facilitate the unpredictable and spontaneous technological and social innovations necessary to circumvent environmental harm and climate change (Prins and Caine 2013). One approach suggests that it is fair to assign responsibility to occupants of institutionalised roles for failing to nurture the social, political, and institutional basis of democratic climate change and for allowing this basis to be exhausted. This is a particular and inappropriate approach if democracy is the only realistic breeding ground for the social and political innovation that is necessary for satisfactory climate policies.

I embrace a so-called *teleological* and *maximal* theory of democracy, and suggest that it is fair to hold the relevant occupants of institutional positions not merely blameworthy for not doing enough to mitigate climate change but also responsible for choosing a wrong approach to climate politics. When such agents focus primarily on the economic incentives aspects of climate politics, the preconditions for providing successful climate politics, which take their point of departure from people's moral inclinations, are undermined. By so arguing, we allow for a new type of critique of the economic approach to climate change, one that fails to recognise that a precondition for successful climate politics is to improve democratic quality.

Conclusion

In this chapter, the fact-sensitive account of moral responsibility was introduced by focusing in particular on the normative criteria of climate actions and politics. This account provides the theoretical platform for much of what follows in Part 3. At the core, the fact-sensitive account of moral responsibility is the criterion of moral blame or excuse and the normative criterion of justification of the assigning of responsibility to particular agents. Other key theoretical notions introduced here are the distinction between individual, joint, and collective responsibility and moreover, the distinction between *the tragedy of the commons* and *the tragedy of the few*, all of which have implications for whether some, if any, are responsible for enhancing the fitness-conditions of democratic climate change.

Notes

1 There are also several terminological differences. Some defend collective responsibility (Feinberg 1968; V. Held 1970). Others defend shared responsibility (Pogge 1992), joint responsibility (S. Miller 2006) and national responsibility (D. Miller 2007). Yet others defend mutual benefit responsibility (Rawls 1971) and assigned responsibility (Goodin 1995).
2 A good example is Miller's account of national responsibility (D. Miller 2007).

References

Feinberg, J., 1968. Collective responsibility. *Journal of Philosophy*, 65 (21), 674–688.
Geuss, R., 2008. *Philosophy and real politics*. Princeton, NJ: Princeton University Press.
Goodin, R., 1995. *Utilitarianism as a public philosophy*. Cambridge: Cambridge University Press.
Held, V., 1970. Can a random collection of individuals be responsible? *Journal of Philosophy*, 67 (14), 471–481.
Kamm, F. M., 2007. *Intricate ethics: rights, responsibilities, and permissible harm*. Oxford: Oxford University Press.
Lichtenberg, J., 2010. Negative duties, positive duties. *Ethics*, 120 (April), 557–578.
Miller, D., 2007. *National responsibility and global justice*. Oxford: Oxford University Press.
Miller, S., 2006. Collective moral responsibility: an individualist account. *Midwest Studies in Philosophy*, 30 (1), 176–193.
Miller, S., 2010. *The moral foundations of social institutions: a philosophical study*. Cambridge: Cambridge University Press.
Miller, S., 2011. Collective responsibility, epistemic action and climate change. *In:* N. Vincent, I. van de Poel, and J. van den Hoven, eds. *Moral responsibility: beyond free will and determinism*. Dordrecht: Springer, 219–246.
Pogge, T., 1992. Cosmopolitanism and sovereignty. *Ethics*, 103 (1), 48–75.
Prins, G. and Caine, M. E., 2013. *The vital spark. Innovating clean and affordable energy for all*. The Third Hartwell Paper, July. London: LSE Academic Publishing.
Rawls, J., 1971. *A theory of justice*. Cambridge, MA: The Belknap Press of Harvard University Press.
van de Poel, I., 2011. The relation between forward-looking and backward-looking responsibility. *In:* N. Vincent, I. van de Poel, and J. van den Hoven, eds. *Moral responsibility: beyond free will and determinism*. Dordrecht: Springer, 37–52.
Vanderheiden, S., 2011. Climate change and collective responsibility. *In:* N. Vincent, I. van de Poel, and J. van den Hoven, eds. *Moral responsibility: beyond free will and determinism*. Dordrecht: Springer, 201–218.

8   Moral excuse and democratic citizens

In this chapter, I consider various arguments for why democratic citizens ought to be held morally responsible for climate change or whether we have reason to excuse them. To do so, I make use of the fact-sensitive account of ought-assignment elaborated on in Chapters 3 and 7. My concern is to articulate the trade-off between demanding too much of democratic citizens and maintaining a liberal-democratic framework in which we have normative reasons to think that individuals should be free to do what they want (as long as they do not infringe upon anyone's rights).

Two preliminary assumptions should be noted. First, it is presumed that the democratic citizens in question meet no hard constraints and qualify arguably, as moral agents (van de Poel et al. 2012, 55). Second, it is assumed that democratic citizens are polluters who contribute (to some extent) to environmental harm and climate change by, for example, refraining from acting in climate-friendly ways – e.g., recycling, turning down the thermostat, driving a car instead of using public transport or cycling, and so forth (Fahlquist 2009, 115). The pertinent question is whether citizens should also be held responsible for the harmful effects of their lifestyle choices and behaviour on the earth's climate and atmosphere, a question that depends on a fundamental principle about whom it may be fair to assign responsibility to.

Two lines of arguments are discussed in this chapter. The first concerns the soft *internal* constraint that democratic citizens do not experience their individual failure to act as failures: people do not psychologically 'feel' they are doing anything wrong (Lichtenberg 2010, 561–562). The second concerns the soft *external* constraint that highlights a number of studies have shown that many consumers are concerned about environmental issues but struggle to translate this into green choices (Thøgersen and Schrader 2012, 1–5; Young et al., 2010, 20–31). Environmental harm caused by modern consumers is the result of a gap between a citizen's green attitudes and their non-green behaviour. Thus, whilst democratic citizens may think they are morally obliged to avoid harming the climate, they struggle to fulfil these obligations, nonetheless.

The chapter proceeds as follows: first, I discuss two reasons for excusing democratic citizens for not fulfilling their moral obligations. I conclude that whilst both arguments fail, they nonetheless still explore why people find it difficult to avoid contributing to climate change and environmental harm. Second, I discuss the extent to which the two analytical distinctions between negative and positive duties and between acts and failures can provide reasons for excusing the group of democratic citizens. I show that this is the case with regard to climate change, but not environmental harm.

The moral deficit and democratic citizens

The purpose of the next sections is to consider various strategies for how the fitness-conditions for responsibility may feed into the discussion of whether democratic citizens should be held morally responsible or be morally excused for climate change. The first strategy proposes that it is unfair to blame democratic citizens for the harm caused by their choices and behaviour if they did not know they were harming the climate (or future generations). This argument presumes a *knowledge deficit*: democratic citizens might not know the environmental impact of their choices and behaviour. This assumption can be motivated by the conclusions in the previous chapters about rational estimation failures. Indeed, the full consequences and net effect of, for example, driving an environmentally unfriendly car, remain difficult to predict. Another type of reasoning assumes that democratic citizens generally know the consequences of their choices and actions but lack the incentive to act accordingly due to, for example, the cost and availability of environmentally friendly products. Here, one may talk about an *incentive deficit*.[1] Both deficits are supported by empirical research on green consumer behaviour. In short, the key issues explaining the non-green consumer choices are (1) the lack of time for research and lack of information, and (2) the availability and the price of green products (Young et al., 2010, 27). They reflect the second fitness-condition of responsibility – namely the question of whether sufficient knowledge is available for the agents. In other words,

> *A1. The moral deficit argument*: (1) Democratic citizens do not know the environmental impact of their choices and behaviour and thus do not know what they should do in order to make green choices. (2) Even if (1) is false, democratic citizens lack an incentive to make green choices because of the prices and the availability of environmentally friendly products. (3) If (1) or (2), then democratic citizens can be morally excused for non-green choices. (4) Thus, democratic citizens can be excused for non-green choices.

One problem with this argument is that (3) it is implausible as a principle for 'whom it is fair to morally excuse'. It could be the case that people ought to know what environmental impact their choices have even if they do not, that is, agents can only be excused for what some call 'non-culpable ignorance' (van de Poel et al. 2012, 54). Likewise, Shue argued that people cannot be excused for something they could have been expected to know (Shue 2006, 3). Thus, the relevant question is, What can we expect democratic citizens to know? For example, Shue argued,

One must be forgiven for not knowing 99.9 percent of what is knowable. The costs of acquiring and the sheer impossibility of acquiring more than a little information are a fairly obvious reason for not holding individual persons responsible for all the havoc they wreak through lack of well-informed foresight.

(Shue 2006, 3)

However, if we presume that the citizens are living in a well-functioning democracy, a reasonable assumption could be that the citizens have access to the relevant knowledge, as argued in Chapter 5. Thus, the moral deficit argument fails because people can be expected to know more or to pursue the available knowledge about, for example, a product's long-term consequences and thereby be motivated to pay more for more environmentally friendly products. In other words, individual agents cannot be excused for not knowing enough about products that cause long-term environmental harm if they have a *first-order responsibility* to seek the knowledge, which we assume for the sake of argument is available in the first place.

Another question is whether democratic citizens can translate their knowledge into concrete actions. This challenge concerns the soft *external* conditions that might constrain the citizens' actions and choices. One challenge is that several vital energy resources and public services cannot be said to be in the direct control of individuals in a meaningful sense. Take, for example, the heating system in a local school. Because the parents may know that the heating system is not based on a renewable energy resource, this may give us reason to claim that they have an obligation to change the system. In the empirical literature, this is called a *locked-in* syndrome, which is when high-income groups or countries are locked into an environmentally and climate unfriendly lifestyle because of a non-green public infrastructure (Csutora 2011).[2] Let us call this a *no-impact challenge*. This challenge reflects the third fitness-conditions of responsibility and questions whether the agent has control over the relevant choices that are available.

Addressing the same challenge, Sinnott-Armstrong has argued that individual responsibility for driving a non-green car does not matter because of the imperceptible harm this single car may cause (Sinnott-Armstrong 2010). If the long-term environmental impacts of green choices and behaviour are only marginally better than non-green choices and behaviour, we may conclude that it is beyond the individual's capacity to change the ecologically destructive consumption that society causes. The latter argument is slightly different from the no-impact argument; let us call it a *low-capacity challenge*.

The alternative moral deficit argument looks like the following:

A2. The moral deficit argument: (1) Democratic citizens' pro-environmental behaviour does not have a green impact. (2) Even if (1) is false, democratic citizens (conducting daily activities) do not have the capacity to make a huge green impact. (3) If (1) or (2), then democratic citizens can be morally excused for their non-green choices and actions. (4) Thus, democratic citizens can be excused for their non-green choices and actions.

Several things should be noted. First, we may question whether it is correct that people cannot make a difference. After all, it seems reasonable to presume that every individual contribution will lead to at least a tiny incremental effect. Nonetheless, we may argue that the comparative effect of green choices in some cases is very small. This does not, however, seem to provide a strong case for a moral excuse.

Another more challenging argument for excusing individual agent's moral deficit concerns an individual's choice to live in climate-friendly or unfriendly behavioural space. Consider a case of public transportation. If Peter is living in an area without public transportation, it would be very difficult not to have a car. In order to reduce his ecological footprint, Peter should move to another area or country with a different institutional behavioural space – i.e., a climate-friendlier public infrastructure. In this case, it may be impossible to blame Peter for using his car because the surrounding behavioural space leaves him no other options.

Another question is whether we can blame Peter for not moving to an environmentally friendly behavioural space. This is a question of how much we can demand from individual agents within, let us say, a broad liberal theoretical framework. In the next section, I discuss this issue as a question of to what extent governmental authorities may force individual agents to move to another city or country. In most liberal views, nobody is allowed to force individual agents to move. The argument is especially strong within a right-libertarian view, which emphasises people's negative rights to self-ownership and physical integrity. Let us, therefore, consider the strongest argument against moving people without their consent – namely, what we could call the right-libertarian moral deficit argument.

The right-libertarian moral deficit argument

Having argued that lack of knowledge is not a sufficient excuse for contributing to long-term climate change and that the no-impact argument would require people to move to a climate-friendlier community, let us now discuss whether democratic citizens can be forced to do something to which they do not voluntarily consent. This challenge reflects the first fitness-condition for moral responsibility, which concerns the question of whether agents face a normatively significant choice. It can be argued that if the agent is acting under compulsion, it is not reasonable to hold her responsible (van de Poel et al. 2012, 54). The relevant question is how we weigh agents' voluntarism in contrast to politics that favour the common good. Let us start with a variant of the argument that puts emphasis on the moral significance of voluntarism.

> B. *The right-libertarian moral deficit argument*: (1) Democratic citizens cannot be forced to do something to which they are not voluntarily committed. (2) Avoiding climate harm requires more than voluntary commitments. (3) If (1) or (2), democratic citizens can be excused for their non-green choices. (4) Thus, democratic citizens can be excused for their non-green choices.

In order to accept this argument, two assumptions need to be accepted. First, the libertarian argument presumes the relevance of negative liberty – i.e., the right to be left to the rule of your own private will (Berlin 1958, 7; Nozick 1974, 160–163). The concept of negative liberty presupposes the distinction between negative and positive duties. In contemporary philosophy, the distinction is widely accepted and is at the core of the debates on global justice and environmental ethics, although several scholars have criticised its moral and conceptual significance.[3] Negative duties are generally defined as the duty not to harm others, whereas positive duties are duties to render someone aid. Distinguishing negative from positive duties allows one to negatively delineate people's moral obligations. The moral significance of distinguishing the two sets of duties is normally considered to be a question of whether the agent has to 'act' or 'refrain from acting'. As long as people are not required to do something but are only asked to refrain from doing something, negative moral duties are believed to be consistent with people's negative liberty.

This leads to the second presumption of the libertarian argument – namely, the proposition that people's negative duties are more stringent than positive duties (Pettit 1997, 17–18; Pogge 2010; Ashford 2009, 93, 101). The basic idea is that as long as people refrain from harming other people, they have fulfilled their duties. Agents should not be burdened with duties that exceed their negative rights not to be violated or interfered with in their private life.[4] If these two presumptions are accepted, one should accept the libertarian moral deficit argument that people care less for the environment and future generations than they care for their private welfare and well-being.

To evaluate the plausibility of this argument, let us again consider the example of the infrastructure-deprived society. Because Peter lives in a behavioural space that has no public transportation, the cost of not having a car is very high. The cost is especially high if schools, supermarkets, and offices are placed in remote places (under the presumption that people have cars). In order to reduce his ecological footprint, Peter would be required to move to another location with a climate-friendly public infrastructure. If we presume that Peter's climate harm is primarily caused by CO_2-emitting car driving, and the only way he could lower his environmental impact would be to move to a more developed public infrastructure area, libertarian reasoning argues that we cannot blame him for his climate impact. Jessica Fahlquist made a similar point. Addressing the possible obstacles of lack of money and infrastructure, (which is assumed to be the result of many years of political decisions), culture, and economic factors, she argued that it does not make sense to blame democratic citizens since the context of alternatives does not provide the conditions for morally blameable actions (Fahlquist 2009, 112–123).

There are, however, two questions going on here that we need to disentangle. One concerns what democratic citizens are responsible for in the context of a lack of climate-friendly infrastructure. The other concerns whether the democratic citizens are allowed to resist a request from, let us say, a legitimate government that is about to build a better climate infrastructure that requires that Peter and other people move to another location. A proponent of right-libertarianism would most likely embrace both, whilst an opponent of right-libertarianism will accept the first statement by morally excusing democratic citizens if the societal and institutional climate-friendly infrastructure is absent, but reject the latter statement. I tend to embrace the second option. Notwithstanding, let us continue the discussion of the libertarian argument to see if it provides a reasonable argument for the first option.

Libertarian reasoning argues that a government is not allowed to coerce Peter to move because we have strong moral obligations not to infringe on his *negative* right and freedom to decide where he wants to live. Hence, distinguishing negative from positive duties allows us to accept the exclusion of positive duties in order to protect the environment. If the agent has a negative right not to be burdened with costly and time-consuming positive duties (to protect the climate) that conflict with the will of the agent, the distinction between negative and positive duties provides the reason for not blaming that the agent for contributing to climate change.

The reason for this conclusion is that the libertarian version of the distinction between negative and positive duties and rights ranks the protection of the agent's negative right to self-authority higher than the agent's positive duties to protect the climate. However, before we accept the libertarian version of the moral deficit argument, we may question the soundness of the reasons for which we would embrace the distinction between negative and positive duties and the claim that negative duties are more stringent than positive ones. If these two points are not accepted, the libertarian argument will fail.

Moreover, if we reject that negative duties are more stringent than positive ones, then we can argue that governmental requests that aim to produce better environmental infrastructure may not infringe upon an agent's negative rights. This reflects the conclusion of Chapter 5, where I argue that the policy options that solve the problems of *the tragedy of the few* do not infringe upon anybody's property rights.

The *negative duties are not enough* argument

Two points that critique the asymmetry between positive and negative duties should be considered. First, one may critique that negative duties are considered to be more stringent than positive duties, which addresses the plausibility of premise B(2): *to avoid climate harm requires more than voluntary commitments*. One claim may be that in order to avoid long-term climate harm, negative duties are insufficient: positive duties are also required. Let us call this argument the *negative duties are not enough* argument. It goes like this: negative duties demand that individual agents are morally obliged to avoid harming other people. However, in the modern global world, fulfilment of what negative duties demand also requires positive duties. Gilabert has succinctly summed up the point in relation to global poverty, and the same point applies for climate change:

> Satisfying negative duties is not enough for securing the absence of global poverty even if all of presently existing poverty can be traced back to the

impact of harmful policies by the global rich. [. . .] A negative duty to compensate for previous harm done may suffice to justify a redistribution of resources compensating present victims at a certain time t1. But given unavoidable differences in natural and social endowments, it is only to be expected that some people will, at time t2, be unable to stay afloat in the new social framework if it does not include a permanent and enforceable positive duty to help those in need.

(Gilabert 2004, 547–548)

The point is not that negative duties are wrongly conceptualised, as we shall see in the second critique, but that they provide an insufficient reason for global and climate justice. Negative duties are not more stringent than positive duties because the moral objectives of negative duties cannot be fulfilled by negative duties alone: as Gilabert argued, positive duties are required as well. Lichtenberg agrees with Gilabert's critique of giving priority to negative duties; he argued in a recent article that 'negative duties – duties not to harm – are more demanding than has usually been thought, and in this respect, they resemble positive duties to render aid' (Lichtenberg 2010, 575). To avoid global harm is so demanding that positive duties are required as well. This means that positive actions are needed – i.e., the global rich and the current generation have to comply with a set of new rules and adapt to changes in patterns of behaviour.

Libertarians generally consider behavioural changes and the need for compliance to be a cost that may infringe people's negative rights to self-determination. Lichtenberg, however, argued,

Much avoidable suffering in the world could be remedied without great cost to those who would have to act or refrain from acting. [. . .] A crucial condition of keeping the costs – whether material or psychic – to individuals low is that they act, or refrain from acting, as part of a collective effort rather than as isolated individuals. Acting collectively diminishes costs for individuals in several ways. Suppose, for example, a city prohibits the use of plastic bags in supermarkets and chain pharmacies, as San Francisco recently did. The policy immediately relieves the individual of two kinds of effortful action.

(Lichtenberg 2010, 576)

As I noted earlier, social norms and rules are here taken to diminish the overall social costs on an aggregated level. Hence, the cost argument fails. If the *negative duties are not enough argument* is accepted, the libertarian approach to negative duties seems implausible because it does not consider that eradicating global harm requires positive duties as well. Note that this argument may not convince the libertarian.

One important clarification needs to be added. Scholars criticising the distinction between positive and negative rights are primarily interested in what we could call *passive* rights holders. The question of who are the relevant rights holders supplements the question of to whom one may assign positive or negative duties.

Passive rights holders are those who may be infringed upon by anthropogenic climate change and environmental harm. In the literature on global justice and climate ethics, passive rights holders are usually the global poor and future generations whose rights may be infringed on by wealthy countries (Pogge 1992; Nussbaum 1996; O'Neill 2000; Brock and Brighouse 2005). Others, for example, right-libertarian scholars, often draw attention to what we could call *active* rights holders (i.e., duty-bearers), whose negative rights should not be infringed upon.[5] Active rights holders are, for example, the global rich, whose rights to non-interference in their lifestyle and behaviour may be infringed upon by political regulation and moral obligations.[6]

Despite the fact that it seems reasonable to claim that positive duties are as equally morally significant as negative ones, a libertarian can uphold the significance of the libertarian moral deficit argument by addressing the negative rights of the active rights holders. By doing so, they give priority to negative rights not to be coerced by claiming that duty-bearers' negative rights cannot be infringed upon without further justification. This argument may hold even if one accepts that positive duties are necessary in order to give the passive rights holders what they are entitled to. When securing the rights of passive rights holders, negative and positive duties are equally morally significant. Thus, a proponent of libertarianism can reaffirm the significance of the moral deficit argument because it is impermissible to infringe upon the active duty-bearer's negative rights. From a libertarian point of view, there is an asymmetry between what the duty-bearers (i.e., active rights holders) are obliged to do and what the passive rights holders are entitled to.

Negative and positive duties

In this section, I turn to the second critique of the moral deficit argument that may convince the libertarian. This critique rejects the analytical relevance of distinguishing between positive and negative duties. This may be a true test of the libertarian argument. The libertarian argument can accommodate the assumption that positive duties and negative duties are of equal significance. But if the distinction between positive and negative duties is rejected, the libertarian argument fails because the (negative) rights of the active rights holders will not be more stringent than the (positive) duties for the passive rights holders. If this argument fails, the libertarian moral excuse for individual agents' moral deficit in regard to climate change fails as well.

The critique of the distinction between positive and negative duties starts by asking how we know whether a duty is a negative or a positive duty. Let us consider Lichtenberg's plastic bag example again. The prohibition of the use of plastic bags in supermarkets is an example of a positive duty: consumers have to change their behaviour and comply with the legislation by bringing their own bags or carrying their purchases by other means. Yet the plastic bag example may just as easily be formulated in a negative fashion: people should *refrain from using plastic bags*. The positive understanding may be defended by drawing attention to the fact that the city of San Francisco has conducted a positive act: they have deliberately agreed to ban the use of plastic bags. This approach, however, does not solve the problem. Other negative duties such as refraining from violating other people's private property rights are also based on city, state, or federal legislation. In spite of that, many libertarians maintain that private property rights are a negative right and that people have a negative duty not to infringe the property rights of others.

The example of the plastic bags (and private property rights) shows that the distinction between negative and positive duties and rights is frequently arbitrarily constructed, or in the words of Jonathan Bennett, caused by 'unexplained linguistic intuitions' (Bennett 1980, 52). Similarly, Peter Singer argued, 'Whether a rule is appropriately or better expressed in negative or positive terms is purely fortuitous – a function of grammar, idiom, and circumstance. Neither form is always superior' (M. Singer 1965, 103). The claim is that negative and positive rules are analytically equivalent. Hence, 'It makes no difference whether we state the rule in the positive form, "If one says something, then one should tell the truth (to the best of one's knowledge and belief)", or in the negative form, "If one says something, then one ought not to lie"' (M. Singer 1965, 99).

Following Singer, the difference is caused primarily by rhetoric or emphasis. Depending on the context and the social circumstances, linguistic phrases may take on a different meaning. Consider, for example, the phrases 'refraining from using plastic bags' and 'allowing for green choices in the supermarket', which – let us for the sake of the argument assume – identify the same social phenomenon. Whilst the distinction between positive and negative duties may have some practical relevance, it reflects no significant analytical difference: it is merely a question of corresponding ideas that are expressed in different ways.

The asymmetry between acts and omissions

If it is accepted that the distinction between positive and negative duties is redundant, the libertarian version of the moral deficit argument is re-joined. The libertarian argument cannot provide reasons as to why individuals should be morally excused. Libertarians may, however, suggest a counter-argument. One may agree with Singer that it is irrelevant whether social events and political states of affairs are described in positive or negative terms because they reflect neither a significant moral nor conceptual meaning. At the level of individual agency, however, we have reason to believe that it does make a difference. The reason for this is that a non-consequentialist account of moral blame considers it morally significant when people do something (i.e., act) or refrain from doing something (i.e., omission).

Note that the relevant distinction is here not positive versus negative but intended *acts* versus unintended *omissions*. When an agent acts, there are typically some intentions behind the act: the agent aims to achieve some intended goals. The result may include unintended consequences as well: when an agent omits to do something, it may be intended or unintended. The relevant point is that agents may be off the hook if climate change is caused by unintended consequences of acts and by unintended omissions related to innocent daily activities. What does the work here is not whether the harm is caused by intended or unintended acts. That would be a variant of the knowledge deficit argument, which I rejected earlier. It is the difference between omissions and acts that is essential. I argue later that this has implications for whether agents can be morally excused for climate change.

However, one may argue, as Bennett convincingly does, that the differentiation between acts and omissions reflects no relevant moral content. It is not the act itself but its consequences that matter. The reasoning is that non-acts (i.e., omissions) have as many possible causal consequences as acts have (Bennett 1995, 42). If we are interested not in the actual physical movements but in the causal effects of movement versus the causal effects of no movement, the distinction is redundant. Let us consider the plastic bag example again. We may agree that the moral content of the action not to use plastic bags does not consist of the physical movement of using a plastic bag compared to using one's reusable bag. Rather, the relevant moral content consists of the climate effects the different bags may have. Hence, it is morally irrelevant whether the causal effects are the result of an action or a non-action.

This issue touches upon the discussion of the equivalence of killing someone and letting someone die.[7] Most people agree that the spatial and temporal location of harm, such as starvation and environmental pollution, is morally arbitrary. We may feel a difference between experiencing the death of a person as opposed to hearing about a famine in the news. But a psychological explanation of our feelings is (of course) not a moral justification (Rachels 1979, 162). Based on this, James Rachels and others argue for the equivalence thesis – namely, that letting someone die is as bad as killing someone (Rachels 1979, 163). Rachels uses the example of someone who refuses to give a sandwich to a starving child. Let us call this the *sandwich example*. Another much-discussed class of examples includes the *rescue cases* in which an agent fails to rescue a drowning child from a pond (P. Singer 1972, 231). The primary argument in this class of examples is based on the fact that the moral reasons we have for claiming that killing is wrong are the same moral reasons by which we find failing to feed a starving person morally objectionable. Thus, it seems reasonable to claim that spatial and temporal location is morally arbitrary, and that we, therefore, have no reason to assess the two cases morally differently.

Despite the plausibility of this equivalence thesis, several points should be considered. Proponents of the equivalence thesis tend to take a consequentialist approach that determines the wrongness of killing someone in terms of the negative effects for the killed person. It is only if we accept a consequentialist account that we are forced to reject the moral significance of distinguishing killing from letting die. If one is a non-consequentialist moral philosopher paying attention to people's duties or intentions, the distinction between positive and negative acts sustains its relevance. The relevant moral content of carrying a reusable bag reflects exactly the point of interest that draws attention to people's duties or intentions to act in an environmentally friendly manner.

It seems, therefore, that the non-consequentialist moral philosopher has a reasonable argument. If the relevance of an individual agent's intentions is rejected or ignored, the result may be a diluted concept of what is morally impermissible in which seemingly daily activities (such as driving an environmentally unfriendly car and thereby failing to avoid harming the climate) are morally judged in the same manner as an agent deliberately killing someone. When discussing which acts are morally blameworthy, people's intentions should be considered relevant. Despite the fact that the same moral ideas evaluate consequences of omissions and actions, as Rachels argued, we are not allowed to say that we cannot use the same reasons to blame individual people for omissions and actions. At least this argument applies if we reject a consequentialist account of moral wrongdoing and take people's intentions into account, which I argued in Chapter 7 was reasonable at the individual level of agency.

Perfect and imperfect duties

A consequentialist does not stress the causal chain but draws attention to the total outcome, for example, the total number of people being harmed by climate change. Note that although we agree with the consequentialist that the total outcome of a complex network of acts, behaviour, and policies is morally bad, we cannot necessarily accept the inference that killing someone and letting someone die are morally equivalent. The discussion about killing someone and letting someone die concerns what kinds of acts and behaviour individual people are obliged to pursue.

Despite the morally objectionable outcomes, it seems unfair to blame an agent for not avoiding acts that harm future generations if the blameworthy act the person has conducted is one of the examples discussed earlier, such as using plastic bags in the supermarket or driving a climate-unfriendly car. Thus, the equivalence thesis cannot be accepted if it is considered morally significant to distinguish deliberately harming someone from failing to avoid harming someone (i.e., a negative duty) or from failing to aid (i.e., a positive duty). This is because the case of climate harm is much more complex than the sandwich case discussed earlier. In fact, the sandwich case does not help us understand the moral dilemmas of climate change. It is wrong to reduce the morally objectionable challenges that global politics is currently facing to a question of blaming individual agents for conducting morally objectionable acts (such as failing to harm the climate). In the cases of climate change, the causal relationship between the perpetrator and the victim is much more multifaceted than omitting to give a sandwich to someone.

In order to distinguish more clearly the cases similar to the sandwich case from those of climate change, one may distinguish between perfect and imperfect duties.[8] Drawing on a Kantian framework, we can say that it is impermissible to treat humanity as merely a means and not also as an end. This moral imperative can be translated into a maxim defending perfect duties to do X if there is a precise claim against A. This is relevant in the case of killing someone. Agent A has a perfect duty not to kill agent B (because B has a claim against A).

In other contexts such as climate change, the claims against A are more imprecise because there are multiple courses of actions by which to pursue A's moral obligations (i.e., to treat humanity as an end). In such cases, it can be claimed that A has only an imperfect duty to avoid causing climate change by his or her daily activities. The reason is that nobody has a precise claim against A. Notice how imperfect duties are as morally obligatory as perfect duties. The difference refers only to the fact that concrete acts pursuing imperfect duties are less clear and determinate, whereas concrete acts pursuing perfect duties are clear and determinate (Gilabert 2006, 195–196).

In the sandwich case, it is obviously morally impermissible to refuse to hand a sandwich to a starving child. Thus, the relevant duty-bearer has a perfect moral duty to give the sandwich to the starving child. The reason is that the child has a clear claim against the relevant duty-bearer. Likewise, in short-term cases of environmental harm, the duty-bearers are obvious. No one is allowed to or can be morally excused for spilling oil in the sea. In long-term cases of climate change, however, it is less clear which acts are morally impermissible and which are not, and which would be an infringement of the duty-bearer's negative rights. Most people would accept that individuals have moral obligations towards the environment and future generations. One reasonable suggestion could be that everybody is morally obliged to treat future generations as an end and not a means. In the long-term cases of climate change, the moral obligations reflect imperfect duties where the morally permissible acts are imprecise. The moral deficits of individuals are not merely caused by infringement on what is morally impermissible, but by the fact that what it is morally impermissible is unclear in the first place.

If we accept this view of the challenges of climate change, we have reason to excuse democratic citizens for their omissions if they are related to imperfect duties. This is not an argument that imperfect duties are morally permissible or that democratic citizens are not obliged to fulfil their imperfect duties in the same manner as in the case for perfect duties. To accept the moral significance of imperfect duties is only to say that democratic citizens frequently face difficulties fulfilling their imperfect duties due to the uncertainty and imprecise character of those duties, and that these difficulties should be considered morally significant and hence as reasons for morally excusing people for their omissions.

Joint responsibility

Until now, I have investigated the role of fitness-conditions in moral judgements about the moral responsibility of democratic citizens. I suggested that in the case of climate change, democratic citizens are excused if they do not intend to cause climate change. A premise for this is that the distinctions between positive and negative duties on the one hand and acts and omissions on the other are accepted. One decisive argument holds that it is unfair to take a consequentialist approach to people's acts and omissions. Evaluating people's acts and omissions morally, I claimed that we should strive to take their intentions into account. If people do not aim or intend to harm the global climate but their behavioural spaces do not allow them to act otherwise, I argued that they are excused for failing to avoid harming the climate.

Note that the implication of this is that democratic citizens are morally responsible for contributing to acts and societal behaviour if they did know the negative consequences thereof and if doing so would not require a change of residence. In other words, people are responsible for all the cases that resemble the moral dilemma of the sandwich case. Thus, in a complete green society, no one would be morally excused because acting in a climate-unfriendly way would be like failing to give a sandwich to a starving child. Although the current moral dilemmas of climate change, sustainability, and global justice do not resemble the moral dilemma illustrated in the sandwich case, one may argue, furthermore, that people are morally obliged to try to transform the moral dilemma of climate change into a question similar to the sandwich case. One example could be the moral obligation to vote for green parties that presumably would take away the burdens of individuals by, for instance, investing in green technologies, efficient energy systems, and green transport.[9]

This refers to the aforementioned *diachronic* aspect of responsibility, which holds agents responsible for not having got themselves into a position to be able to what is morally required to do (Gilabert and Lawford-Smith 2012, 811). However, as also mentioned earlier, in many cases and in the case of climate change in particular, which has a complex and indirect route between cause and consequence, democratic citizens do not have the ability to put themselves into the 'right' position on climate by themselves. This applies at least if we accept the earlier account of individual agents' lack of capacity to change the environment of climate unfriendliness that still dominates most countries in the world. Thus, I maintain that it is unfair and may defeat the purpose to focus on what the group of democratic citizens should be held responsible for. Instead, it is fair to assign *joint* responsibility to democratic citizens for not having done enough earlier in order to provide the conditions of being able to something now.

Conclusion

In this chapter, I have discussed the normative criteria for whether democratic citizens ought to be held responsible for climate change. I have argued that there are significant moral limits on assigning moral responsibility to single democratic citizens by virtue of their moral rights, which should not be infringed. Finally, I concluded that it is fair to hold democratic citizens as a group jointly responsible for not having done enough to mitigate the effects of climate change at an earlier point in time.

Notes

1 A recent study shows that Danish young people have knowledge about climate change but nonetheless do not act in an environmentally friendly way. See Gundelach et al. (2012).
2 For empirical studies of the attitude-behaviour gap, see Csutora (2011), Jackson (2005), Cohen and Murphy (2001), and Ölander and Thøgersen (1995; 2006).

3 For good exploration of this issue, see Shue (1980, Chapters 1–2). For recent contributions with regard to global poverty and climate change, see Gilabert (2004), Ashford (2009), Lichtenberg (2010), and Beitz and Goodin (2009).
4 For further elaboration on negative freedom as non-inference and non-dominance, see Pettit (2001, 128–129).
5 Right-libertarianism espouses the principle that each agent has a right to equal negative liberty and self-ownership, where negative liberty and self-ownership are the absence of forcible interference from other agents (e.g., Narveson 1988; Narveson and Sterba 2010). Right-libertarianism is to be distinguished from left-libertarianism, which holds that natural resources belong to everyone in some egalitarian manner (see Steiner 1994; Cohen 1995).
6 See Shue (1980, 164–166) for a similar argument. For a recent discussion, see Gosselin (2006).
7 The debate on the equivalence thesis refers to the discussions of abortion, euthanasia, and global poverty. See, e.g., Trammell (1975), Foot (1978), P. Singer (1972), and Rachels (1979, 1986). For a general discussion of this issue, see Kagan (1984). More recent contributions include Lippert-Rasmussen (1998, 2007).
8 Note that there is no consensus on the relevance of distinguishing between perfect and imperfect duties. For at good overview of the discussion, see, e.g., Hill (1992, Chapter 8) and Johnson (2012).
9 For example, Maltais (2013) argued that people have a first-order obligation to vote for green parties.

References

Ashford, E., 2009. The alleged dichotomy between positive and negative rights and duties. *In:* C. R. Beitz and R. E. Goodin, eds. *Global basic rights*. Oxford: Oxford University Press, 94–112.

Beitz, C. R. and Goodin, R. E., eds., 2009. *Global basic rights*. Oxford: Oxford University Press.

Bennett, J., 1980. *Morality and consequences*. The Tanner Lectures on Human Values, Brasenose College, Oxford University, 9, 16, and 23 May.

Bennett, J., 1995. *The act itself*. Oxford: Oxford University Press.

Berlin, I., 1958. *Two concepts of liberty*. Oxford: Oxford University Press.

Brock, G. and Brighouse, H., eds., 2005. *The political philosophy of cosmopolitanism*. Cambridge: Cambridge University Press.

Cohen, G. A., 1995. *Self-ownership, freedom, and equality*. Cambridge: Cambridge University Press.

Csutora, M., 2011. From eco-efficiency to eco-effectiveness? The policy-performance paradox. *Society and Economy*, 33 (1), 161–181.

Fahlquist, J. N., 2009. Moral responsibility for environmental problems – individual or institutional? *Journal of Agricultural Environmental Ethics*, 22 (2), 109–124.

Foot, P., 1978. The problem of abortion and the doctrine of the double effect. *In:* P. Foot, ed. *Virtues and vices in moral philosophy*. Oxford: Clarendon Press, 19–32.

Gilabert, P., 2004. The duty to eradicate global poverty: positive or negative? *Ethical Theory and Moral Practice*, 7 (5), 547–548.

Gilabert, P., 2006. Basic positive duties of justice and Narveson's libertarian challenge. *Southern Journal of Philosophy*, 44 (2), 193–216.

Gilabert, P. and Lawford-Smith, H., 2012. Political feasibility: a conceptual exploration. *Political Studies*, 60 (4), 809–825.

Gosselin, A., 2006. Global poverty and responsibility: identifying the duty-bearers of human rights. *Human Rights Review*, 8 (1), 35–52.

Gundelach, P. et al., 2012. [The social condition of the climate.] *Klimaets sociale tilstand*. Aarhus: Aarhus Universitetsforlag.

Hill, T. E., Jr., 1992. *Dignity and practical reason in Kant's moral theory*. Ithaca, NY: Cornell University Press.

Jackson, T., 2005. Motivating sustainable consumption: a review of evidence on consumer behaviour and behavioural change. *Energy & Environment*, 15 (6), 1027–1051.

Johnson, R., 2012. Kant's moral philosophy. *In:* E. N. Zalta, ed. *The Stanford encyclopedia of philosophy*. Summer 2012 edn. Available from: http://plato.stanford.edu/archives/sum2012/entries/kant-moral/ [1 January 2017].

Kagan, S., 1984. Does consequentialism demand too much? Recent work on the limits of obligation. *Philosophy and Public Affairs*, 13 (3), 239–254.

Lichtenberg, J., 2010. Negative duties, positive duties. *Ethics*, 120 (April), 557–578.

Lippert-Rasmussen, K., 1998. Are killing and letting die morally equivalent? *Danish Yearbook of Philosophy*, 33, 7–29.

Lippert-Rasmussen, K., 2007. Why killing some people is more seriously wrong than killing others. *Ethics*, 117 (4), 716–738.

Maltais, A., 2013. Radically non-ideal climate politics and the obligation to at least vote green. *Environmental Values*, 22 (5), 589–608.

Narveson, J., 1988. *The libertarian idea*. Philadelphia, PA: Temple University Press.

Narveson, J. and Sterba, J. P., 2010. *Are liberty and equality compatible?* New York, NY: Cambridge University Press.

Nozick, R., 1974. *Anarchy, state, and utopia*. New York, NY: Basic Books.

Nussbaum, M., 1996. Patriotism and cosmopolitanism and Reply. *In:* J. Cohen, ed. *For love of country*. Boston, MA: Beacon Press, 3–20.

Ölander, F. and Thøgersen, J., 1995. Understanding of consumer behaviour as a prerequisite for environmental protection. *Journal of Consumer Policy*, 18 (4), 345–385.

Ölander, F. and Thøgersen, J., 2006. The A-B-C of recycling. *European Advances in Consumer Research*, 7, 297–302.

O'Neill, O., 2000. *Bounds of justice*. Cambridge: Cambridge University Press.

Pettit, P., 1997. *Republicanism: a theory of freedom and government*. Oxford: Oxford University Press.

Pettit, P., 2001. *A theory of freedom. From the psychology to the politics of agency*. London: Polity Press.

Pogge, T., 1992. Cosmopolitanism and sovereignty. *Ethics*, 103 (1), 48–75.

Pogge, T., 2010. *Politics as usual: what lies behind the pro-poor rhetoric*. Cambridge: Polity Press).

Rachels, J., 1979. Killing and starving to death. *Philosophy*, 54 (208), 159–171.

Rachels, J., 1986. *The end of life: euthanasia and morality*. Oxford: Oxford University Press.

Shue, H., 1980. *Basic rights: subsistence, affluence, and U.S. foreign policy*. 2nd edn. Princeton, NJ: Princeton University Press.

Shue, H., 2006. *Normative and empirical evaluation of global governance, Princeton standards of accountability: avoiding simplistic domestic analogies*. Department of International Relations, University of Oxford, Oxford, UK, 1–8.

Singer, M. G., 1965. Negative and positive duties. *The Philosophy Quarterly*, 15 (59), 97–103.

Singer, P., 1972. Famine, affluence, and morality. *Philosophy and Public Affairs*, 1 (3), 229–243.

Sinnott-Armstrong, W., 2010. It's not my fault: global warming and individual moral obligations. *In:* S. M. Gardiner, S. Caney, D. Jamieson, and H. Shue, eds. *Climate ethics: essential readings*. Oxford: Oxford University Press, 314–331.

Steiner, H., 1994. *An essay on rights*. Cambridge, MA: Blackwell.

Thøgersen, J. and Schrader, U., 2012. From knowledge to action: new paths towards consumption. *Journal of Consumer Policy*, 35 (1), 1–5.

Trammell, R., 1975. Saving life and taking life. *Journal of Philosophy*, 72 (5), 131–137.

van de Poel, I. et al., 2012. The problem of many hands: climate change as an example. *Science and Engineering Ethics*, 18 (1), 49–67.

Young, W., Hwang, K., McDonald, S., and Oates, C. J., 2010. Sustainable consumption: green consumer behaviour when purchasing products. *Sustainable Development*, 18 (1), 20–31.

9   Collective responsibility and democratic institutions

As we have seen, according to my fact-sensitive account of ought-assignments, the ultimate assignment of moral responsibility to particular agents depends on two normative criteria – that is, that agents are not morally excused and that there is a solid, positive justification for why it is fair to assign moral responsibility to particular agents. By invoking this definition of collective responsibility, loosely connected groups, such democratic citizens, should not be held collectively responsible for climate change because, as I argued earlier, they can – under certain circumstances – be morally excused for things they did not intend. Their wrongdoing consists of seemingly innocent acts such as driving their kids to school and using the electricity available in the city. I now turn to the question of whether it is fair to assign moral responsibility to the *occupants of institutional positions* in modern democracies.

I undertake two main tasks in this chapter. First, I apply my fact-sensitive account of ought-assignments to occupants of institutionalised positions in modern democracies by investigating the importance of addressing the dynamic aspects of collective responsibility and by discussing what occupants of institutional positions should be held collectively responsible for. Second, I elaborate on the normative implications of the account of collective responsibility for whether democratic climate governance should be institutionalised at the national or the global level.

Collective institutional duties

According to the fact-sensitive account of moral responsibility, the first normative condition for responsibility assignment is the moral excuse. At the level of the collective level of agency, this is a conceptual point. I argued earlier that a defining feature of institutionalised group agents is the distinction between input and output sides, and that moral excuse only applies to the input side; arguably, the collective level of agency cannot be morally excused for their deficit with regard to climate governance. I then continue to the second normative condition for assigning responsibility, which is positive justification of assignment of moral responsibility to particular groups of agents. In this chapter, I focus on the occupants of institutionalised positions in modern democracies and elaborate on what exactly it is fair to hold them responsible for.

A robust positive theory says something about why it is that occupants of democratic institutional positions *ought* to be held collectively responsible for climate change. This discussion draws on the OIC principle: *ought implies can*. As argued in Chapter 3, OIC can be approached in several ways. I have focused on the presuppositional and the entailment understandings of OIC and have argued in favour of the latter. An important aspect of the entailment version of the OIC is that the main account of what one agent *can* do is not merely dependent on what she can do at a specific point of time (T2).

In contrast, the account of what one agent can do includes what the agent has done at some earlier point in time (at T1) in order be capable of doing what ought to be done at T2. To reiterate the earlier discussed wedding example: Louis did not show up at his wedding at 9 am because he had boarded a plane to Los Angeles at 8:30 am. According to the entailment understanding of OIC, the important point to stress is that Louis is not primarily blameworthy for not showing up at 9 am because he no longer can. What he is blameworthy for is that he boarded the plane at 8:30 am, which made it impossible for him to show up in Boston at 9 am.

The justification for this interpretation of the OIC is the notion of dynamic duties, which looks at whether moral agents have aimed (at T1) at getting in the right position to be capable of doing what ought to be done (at T2). This account of dynamic duties applies to both the group of democratic citizens and the occupants of institutionalised positions. Although democratic citizens should be morally excused for not doing enough for the climate because they are unable to do something which has a high impact on the climate (at T2), they should be held responsible if they could have done something more at an earlier point in time (at T1). As we have seen, they could, for example, have voted for a green party that would have legislated in favour of climate-friendly policies (at T1) that arguably would have raised the citizens' positive impact on the climate (at T2).

In contrast, occupants of institutionalised positions may be held collectively responsible for not having promoted an infrastructure in which, for example, car driving and electricity use are not morally blameworthy acts. A climate-friendly infrastructure is an example of what I identified earlier as soft external constraints, which are 'necessary enablers' for allowing single agents to do what they ought to do (Howard-Snyder 2006, 239). In this sense, the occupants of institutionalised positions should be held collectively responsible for remedying the negative effects of climate change and for enabling an infrastructure in which it makes sense to hold individual agents individually and jointly responsible for their climate-unfriendly acts and choices. As we shall see, the occupants of institutionalised positions of democracies have not done enough (at T1) with regard to enabling climate-friendly behaviour and policies (at T2), and I argue that they *ought* to be held collectively responsible for that.

Collective ends

We should also consider one further justification of why it is fair to assign responsibility for climate change to institutional position occupants. Miller argued that the purpose of the institutional positions is defined by realising the collective goals and the *raison d'être* of the institution (S. Miller 2010, 54). By this is meant that the institutional roles which individual agents occupy are constitutive of the institutional functions and practices. One way to put it is that every institution, such as a democratic government, a university, and the police, has certain collective goals which constitute the value and purpose of the institution (S. Miller 2010, 225).

The institutionalised positions are defined by the institutional functions and aims that the occupant of the institutionalised positions should execute. In the case of democratic political institutions, institutional position occupants of the democracy, such as public servants in the government and other defining roles, are obliged to fulfil these positions. If they do not do so, the democratic institution may cease to exist and turn into something different from a democratic institution. Thus, we can derive the normative principle that occupants of institutional positions are obliged to fulfil the collective ends of the institutions and thus that it is fair to hold the occupants of these roles collectively responsible if they do not satisfy these institutionalised collective ends.

One reason why the occupants of institutionalised positions are obliged to fulfil the collective ends is because these collective ends define the function and purpose of the institution. Consider one example. One may argue that freedom of speech is constitutive of democratic decision making. Thus, if the occupants of institutionalised democratic positions, for example, rule against the right to freedom of speech, it would be self-contradictory, since democratic decision making would no longer be fully democratic (Rostbøll 2009, chapter 1). With regard to climate politics, however, it is more difficult to argue that one policy framework is self-contradictory because it is not completely clear whether any policy is constitutive of successful climate politics.

I note that the account of which necessary enablers are obligatory differs according to how the problem of climate change is perceived. The different views on the collective goals of political institutions have implications for the assignment of collective responsibility to occupants of institutionalised positions. When we discuss what occupants of institutionalised positions should be held collectively responsible for, one premise is whether we presume that political institutions ought to solve the problems of *the tragedy of the commons*. Or whether we assert instead that occupants of institutionalised positions should be held collectively responsible for the problems of *the tragedy of the few*.

I argued earlier in favour of *the tragedy of the few* as one of the most pertinent challenges of contemporary climate politics. However, this does not imply that *the tragedy of the commons* ceases to be part of a comprehensive climate political package. My point is that the policies aim to solve the incentive problems related to *the tragedy of the commons* are unsatisfactory. Climate politics ought to address the political and normative challenges related to the issues of equal access, benefits, and consumption that *the tragedy of the few* stresses as well. One way of conceptualising the difference between *the tragedy of the* commons and *the tragedy of the few* is to consider whether the collective ends of climate politics that occupants of institutionalised positions are obliged to fulfil have a *nomocratic* or *teleological* property.

The *nomocratic* account takes a minimal approach to the goals of contemporary political and democratic institutions by focusing primarily on policies that set out the rules within which individual agents can fulfil their goals (Plant 2009, 85). Examples of a nomocratic policy are the those related to *the tragedy of the commons*, which aim to mitigate CO_2 emissions in a cost-efficient way and manipulate the rational agents' calculation of costs and benefits with regard to reducing these emissions (OECD 1992; Jagers and Duus-Otterström 2008). The market-based climate policies mentioned earlier can thus be conceptualised as nomocratic policies because the occupants of the institutionalised positions are not defining political goals in any substantive sense. It is up to the agents in the marketplace to fill in what the content of the climate policies should be (Plant 2009, Chapter 1). Within this framework, the occupants of institutionalised positions are only responsible for providing an economic infrastructure that with minimal political means will create the necessary enabling of soft external constraints for climate-friendly behaviour.

In contrast, the *teleological* account includes more substantive goals, in the sense of social justice, distribution, and sustainability as political goals that the occupants of the political institutions are obliged to implement. Good examples of teleological policies are those related to *the tragedy of the few* of how to optimise the benefits of sustainable resources in the air, atmosphere, and water and distribute access to these benefits equally (Ostrom et al. 1994; Schlager et al. 1994; Sagoff 2008, 10). By drawing attention to the importance of governing environmental sustainability and equal access to natural resources politically, the teleological approach provides a more comprehensive account of what climate politics should include. By so arguing, the occupants of institutionalised positions are responsible for providing a more comprehensive infrastructure, including financial, social, technical, and political aspects of public services, that enables climate-friendly behaviour and choices.

In other words, the two different conceptualisations of what collective goals political institutions should pursue reflect divert approaches to climate politics and the questions of *what can be done* and *what ought to be done*. By invoking *the tragedy of the few* and thereby a teleological account of climate politics, I identify in the following discussion two states of affairs for which occupants of institutionalised positions should be held collectively responsible.

Dynamic duties

I argue in this section that occupants of institutional positions should be held collectively responsible for the outcome of a one-sided climate politics that focuses primarily on the problems related to *the tragedy of the commons*, thereby neglecting the problems related to *the tragedy of the few*. I focus on the responsibility for the consequences of what I call a *vicious political circle*. This vicious circle happens when policies are implemented that either can be predicted to be or later are shown to be incapable of solving the failures of climate governance and when alternative policy strategies become less feasible. The vicious political circle of climate politics is a negative side effect of implementing policies that aim to solve ostensible incentive problems. If this is correct, it has implications for how to address the question of the dynamic duties of institutional position occupants.

The vicious circle challenge consists of four elements. Let us examine whether institutional position occupants should be held morally responsible for allowing the vicious circle to happen. As shown in Chapter 6, under the assumption that people do not have the economic incentives to engage in climate-friendly behaviour,

1   Neoliberal and proactionary policies are implemented, manipulating consumers' economic incentives and/or moral inclinations.

Note that whilst neoliberal and proactionary policies are in principle compatible with political and democratic climate governance, they are embraced as alternative solutions to the failures of democratic climate governance. They represent policies that are channelled through mechanisms other than democratic decision making. As we have seen, the two main climate change mitigation designs suggested are the market and authoritarian elitism. If the political problem of climate change is assumed to be one of CO_2 emissions and their attendant costs, technocratic and non-democratic policies might be the best conductor of efficient climate solutions by stimulating, for example, investments in green technologies and the mitigation of CO_2 emissions.

Research, however, shows that a negative side effect of the emphasis on technocratic and elitist policies is the weakening of countries' institutional capacity to facilitate, monitor, and coordinate common problems (Fukuyama 2004, 119). Whilst the budgetary cut-back of the size of states in many countries has been for the good, economic globalisation has tended to erode 'the autonomy of sovereign nation-states by increasing the mobility of information, capital, and, to a lesser extent, labour' (Fukuyama 2004, 119). This leads to the second component of the vicious circle challenge:

2   Neoliberal and proactionary policies risk lowering the social and political capacity to process other problems related to climate change.

As I demonstrated earlier, neoliberal and proactionary policy solutions presume the existence of a stable and benevolent political system. One critical observation, however, can reasonably question whether this stable and benevolent political system can be sustained within the policy design that the neoliberal, proactionary, and authoritarian regimes suggest. How are the problem-solving capacities of the institutional design thought to be maintained? There is a risk that by advancing neoliberal policies and proactionary experiments with human nature, the feasibility of the current political society to process climate policies, including neoliberal policies and proactionary experiments, will gradually be undermined.

Recall that democratic climate governance is dependent on independent and competent citizens who together empower the capacity of democratic institutions to govern climate challenges. If the institutional, social, and epistemic capacity of democratic institutions are undermined, the possibility that democratic institutions can satisfy the fitness-conditions for collective responsibility is undermined too.

There is a growing concern among social scientific scholars that the capacity of current political institutions to cope with difficult political challenges, such as environmental regulation, and other types of challenges, such as right-wing radicalism, social inequality, immigration, and financial insecurity, is already beginning to deteriorate (Scharpf 2006; Plant 2009; Peck 2010; Prins and Caine 2013). Whilst international neoliberal policies have extended economic sovereignty beyond the nation-states, they have simultaneously, Plant argued,

> eroded the powers formerly available to state institutions to correct political and economic imbalances resulting from the operations of the market within their own borders. Indeed, many of the rules agreed under free trade agreements effectively prevent states from unilaterally adopting progressive social, economic or environmental legislation.
>
> (Plant 2009, 14)

Other critiques suggest that by enforcing market-based policies, 'a systematic problem-solving gap' in the political arenas is very likely (Scharpf 2006, 56). Moreover, Fukuyama argued,

> While we do not want to return to a world of clashing great powers, we do need to be mindful of the need of power. What only states and states alone are able to do is aggregate and purposefully deploy legitimate power. This power is necessary to enforce a rule of law domestically, and it is necessary to preserve world order internationally [. . .]. What has de facto filled that gap is a motley collection of multinational corporations, nongovernmental organizations, international organizations, crime syndicates, terrorist groups, and so forth that may have some degree of power or some degree of legitimacy but seldom both at the same time.
>
> (Fukuyama 2004, 120–121)

If we accept that technocratic and elitist policies lower the capacity of the institutional political design by being less capable of reaping the benefits of democratic governance, then we should be aware of another challenge. If governmental authorities, economists, and public thought leaders are assuming that the political problem of climate change is only to manipulate citizens' motivation to mitigate costly CO_2 emissions, the failures of climate governance caused by deteriorating democratic quality and institutional capacity will not be understood as institutional failures. Such failures may be interpreted instead as a persistent motivation and incentive problem that provides sustained support for the neoliberal and proactionary critique of democratic climate governance. One reason for this is that the theoretical critique of the deficiencies of the cost and motivation approach only is difficult to see in the sociopolitical realm. This means that empirical failures are not a sufficient criterion of true evidence because one theory's proof might be another's irrelevance (Blyth 2013, 204 205).

The framework of neoliberal and proactionary policies might be resistant to most empirical factors that prove that the understanding of the problem is wrong. The resistance is caused by the reductionist approach to methodological assumptions about agency and the particular understanding of the problems of climate change that make the proponents of the framework unable to interpret these empirical factors as nothing but evidence that more needs to be done to solve the motivation and incentive problem problems. Hence, the next two steps in the political vicious circle are as follows:

3   The declining institutional capacity of democracies will support the neo-liberal and proactionary critiques of the failures of democratic institutions, increasing the likelihood for further neoliberal and proactionary initiatives.
4   Fulfilling the prophecy, this increases the necessity for tougher anti-democratic authoritarian interventions.

The vicious political circle of climate politics occurs when a political challenge is attacked with the wrong instruments. The challenge of the lack of incentive to engage in climate action might seem hard to overcome without weakening the policy control of political institutions. This approach, however, will only aggravate the situation if the correlation between the lack of incentives and the lack of climate action is spurious. If both the lack of incentives and the lack of climate action are caused by a third explanatory variable – namely, the design and capacity of political institutions and therefore a problem-solving deficit, the wrong policy answers have been correlated with the wrong conceptualisation of the problem. If this vicious circle is not stopped, the democratic mismanagement of climate change will become a self-fulfilling prophecy, and the capacity of the political institutions of democracy to address the societal problems related to climate change will continue to diminish.

To conclude my account of fact-sensitive moral responsibility, it is fair that occupants of institutionalised positions be held morally responsible for not fulfilling their dynamic duties with regard to sustaining the capacity of democratic institutions to provide the collective ends related to climate governance. A further implication of this is that it eliminates the possibility of individual agents to be moral agents with regard to climate change. As long as the occupants of institutionalised positions do not fulfil their responsibility to make the societal infrastructure climate-friendly, it follows that democratic citizens will be less capable of making climate-friendly decisions.

As we have seen, democratic citizens can, of course, do several things to improve the environment, and arguably, also climate change. These contributions might be imperceptible but are nonetheless morally significant. Nevertheless, if the general economic, industrial, and technological environment does not favour climate-friendly behaviour, it is unfair to blame the democratic citizens. I have argued in this section that it is fairer to assign ought-requirements to those who are responsible for promoting a one-sided climate policy framework that has resulted in an economic, industrial, and technological environment which does not support climate-friendly behaviour.

Climate science responsibility

Note that the problem of one-sidedness in climate politics is not merely one of politics but also one of scientific research. A scientometric study of IPCC Third Assessment Report provides evidence of a physical and economic bias in climate change research (Bjurström and Polk 2011). Whereas the natural sciences are dominated by the earth sciences, the social sciences are dominated by economics. A consequence of economically dominated social scientific climate studies is the under-theorisation of the capabilities of the world's political societies to curb climate change politically and collectively. As we saw earlier, one cause of this under-theorisation is the methodology of rationalistic individualism.

We have established that individualism is a solid moral theory for agency. But if it is combined with economic-rationalism, it provides a too narrow an understanding of how political problems can be perceived and addressed. When rational-individualism is used as a political theoretical framework for how to design climate policies, for example, one implication is that the societal and political problems of environmental harm and climate change are depoliticised, de-institutionalised, and de-contextualised. This is because the economic rationale provides no concept of moral agents at the institutional level of agency and, arguably, no normative theory for these institutional agents and practices, as argued in Chapter 4. This problem has been summarised cogently by Nico Stehr:

> By focusing on the effects or goals of political action rather than its conditions, the contentious issue of climate change is reduced from a sociopolitical to a technical issue. The result of these considerations is the depoliticization of climate change (and the politicization of climate science). By concentrating on the effects that require mitigation efforts, the impression is left that the remedies are primarily subject to technological regulation and adjustments. These factors – including the reduction of a sociopolitical to a technical issue – have led among some prominent voices in the community of climate scientists and the science of climate policy to a now clearly discernible skeptical attitude towards democracy.
>
> (Stehr 2013, 58)

In the same manner as incentive-based policies may undermine the influx of moral and social attitudes into the political system, the economic focus on climate change may have the unfortunate consequence of crowding out the relevance of the political and institutional levels of analysis. Some have called this disregard a 'Faustian bargain' – i.e., environmentalists and climate ethicists 'sell' their theories of political agency and ethical beliefs for a few powerful ideas on how to economically value greenhouse gas emissions (Sagoff 2008, 15). Stehr and Storch argued similarly:

> Concentrating climate policy on the reduction of greenhouse gases serves no purpose, if it leads at the same time to preventing taking precautions in dealing with present dangers and their possible future amplifications. Such a one-sided research perspective and climate protection policy will neither protect the climate from society in the coming decades, nor society from the climate.
>
> (Stehr and Storch 2009, 57)

Note that an important component of implementing satisfactory climate policies that also address the problems of equal distribution of access and benefits related to *the tragedy of the few* is that all relevant agents in the democratic decision-process have access to the necessary information. I have stressed earlier the importance of free media, free education, and strong transparency of political decisions. Of particular importance is knowledge dissemination from climate scientists. If the social climate scientists focus too narrowly on the problems related to *the tragedy of the commons*, a critical approach to the problems related to *the tragedy of the few* remains neglected. Thus, I conclude that there is a need to shift or least supplement the currently dominant methodological framework for climate science and climate politics.

This discussion has presented two ways of comprehending new ways of analysing the challenges of climate politics and carbon pollution, one which has redefined the core problems of climate change from being primarily that of *the tragedy of the commons* to that of *the tragedy of the few*. Moreover, by suggesting a fact-sensitive approach to political theory, I have tried to overcome the disciplinary silos between theory and empirical research and allow for a new integration of the different approaches to climate studies.[1]

This research strategy is advanced by the non-reductionist approach, which allows for simultaneous analyses at different levels of agency. A multi-level approach is needed that is capable of addressing the current social and political challenges at various levels of agency. Without discarding issues relating to individual agents' lack of motivation and incentives, non-reductionist institutionalism provides a suitable framework for studying political activities at the collective level of agency and responsibility.

The reason for this shift is that the relevant explanatory level may shift depending on the research question. This approach is inspired by *stance*-theory for the natural sciences (List and Pettit 2011, 13). For example, if one is interested in understanding the biology of an organ, a biological stance may be appropriated. If one is interested in understanding the neurological mechanism of a brain cell, a neurological stance may be appropriated. Likewise, within the social sciences, if one is interested in understanding the motivations or intentions of one single act, an individualist stance may be appropriate. If one is interested in understanding the social coordination of climate governance, a collective and institutional stance is appropriate.

By approaching climate studies in this way, the institutionalist approach opens the political toolbox and provides a range of different policy instruments. This is necessary if the specific normative and factual circumstances of environmental harm and climate change should be addressed. Furthermore, as we have seen, this approach allows for a re-interpretation of the necessity of enhancing democratic decision making and the need for democratic reforms as a way to empower the institutional capacity of effective climate politics.

By so arguing, I conclude that it is also fair to assign collective responsibility to natural and social scientists who accept too narrow a focus on the economic costs and benefits of climate mitigation policies. As a collective group, they have not fulfilled their role in modern democracies to provide the scientific basis and intellectual impetus for the technological and social innovation necessary for understanding the social and institutional dynamics, opportunities, and obstacles with regard to climate politics.

The teleological account of democracy

I have argued that occupants of institutional positions in democracies ought to be held responsible for the negative effects of the vicious political circle of implementing one-sided climate policies. I now turn to the second state of affairs for which occupants of institutional positions ought to be held responsible – that is, solving the neglected problems of climate governance – namely, the equal access and benefits problems related to *the tragedy of the few*. In Part 2, I argued that it is only democratic institutions that are fit to solve these problems. Thus, to answer the question of who ought to solve these problems, we need to answer at the same time at what institutional level democratic climate governance should be institutionalised.

Furthermore, I note that according to the fact-sensitive account of normative principles for institutional practices elaborated on in Chapter 2, we cannot without further justification prioritise the global to the national level or the national to the local level. Let us say a little bit more about the conceptualisation of democratic theory we are engaging here. It is important to distinguish the *intrinsic* and *instrumental* justifications of democracy. The intrinsic justification points to democratic participation and deliberation as the only venues for realising individual's right to liberty. Indeed, some argue that individuals' freedom is not the 'end' of democracy, but it is 'what democracy *is*' and 'a form of exercising freedom' (Rostbøll 2009, 43). In contrast, when we invoke the instrumental justification of democracy, we draw attention to the collective ends that democratic institutions are capable of pursuing.

The instrumental justification of democratic institutions is an important aspect of comprehending the value of democracies in climate governance. As noted earlier in this chapter, it is convenient to approach collective ends in a *nomocratic* or *teleological* order. Recall that the nomocratic approach provides a minimal definition of what the collective ends of democratic institutions are, whereas the teleological approach takes a maximal account of what the occupants of institutionalised positions are obliged to fulfil. Let us now use the distinction between nomocratic and teleological orders to elaborate on the instrumental value of democratic institutions.

Roughly speaking, democratic theories vary according to a continuum between *minimal* and *maximal* concepts. The minimal democracy is a democracy with few formal procedures, such as one election every four years, whereas the maximal democracy includes civil society, the public sphere, and strong participatory and deliberative components that supplement the formal procedure of elections (Rostbøll 2009, Chapter 1). The minimal concepts would take democracy to be compatible with elitist climate policies since public participation and inclusion is only expected every fourth year; in the meantime, political governance should be technocratic and elitist (Rostbøll 2009, 22). In contrast, when we embrace the maximal account of democratic institutions, climate policies that are technocratic and elitist violate the fundamental collective ends of democratic institutions.

Collective ends and democracy

My fact-sensitive account of normative principles is not intended to offer an ideal theory of democracy and global institutions, but to suggest a set of concrete normative criteria for what institutions would be preferable given the issue at stake and the factual circumstances of democratic politics. The concrete normative criteria are issue-specific and relate in this context to the problems of *the tragedy of the few* which ought to be managed. Recall that *the tragedy of the few* story conceives environmental harm and climate change as the result of unequal access to natural resources. I consider in this section whether the instrumental and teleological account of the collective ends related to *the tragedy of the few* provides normative features that allow us to conclude something about at what level of institutionalised politics should be enhanced.

In the instrumental and teleological account of collective ends, the democratic quality of political institutions should be improved by safeguarding, among other things, what in previous chapters has been established as essential for democratic climate governance. These are (1) epistemic gains from collecting information from independent and competent citizens and (2) democratic control of access to consumption and benefits of natural resources. Therefore, climate politics and natural resource management should be distributed to as many people as possible in order to avoid the inevitable wrongness of elitist and centralised decision making.

I argued earlier that well-functioning democratic institutions satisfy the conditions of knowledge and control necessary to conduct policies that distribute the access equally. Now the pertinent question is at what level of governance should the democratic institutions be strengthened. In assessing what level democratic institutions should be improved, I subdivide equal access policies to (1) equal access to decision making about natural resources, (2) equal access to consumption of the resources, and (3) equal distribution of the benefits of the resources (Kaul and Mendoza 2002, 102). By employing this set of criteria, the discussion of democratic and equal inclusion in decision making and the equal rights to consumption of vital resources is supplemented with a fundamental mechanism of democratic governance – namely, equal distribution of benefits related to the natural resources. The question of benefits concerns in these contexts the revenues of selling the resources.

The point I want to pursue is the following: equal access to consumption can be safeguarded in many ways and through multiple institutional designs. Access to consumption is a question of *output legitimacy* (Uhlin 2010, 24). If the goal is merely to safeguard equal access to consumption of vital resources, international institutions promoting effective neoliberal or proactionary climate politics, for example, might be as legitimate as national institutions.

In contrast, if we are more ambitious with regard about safeguarding not merely equal consumption but also equal access to decision making about how to govern the equal consumption of vital resources (Caney 2011, 507), then it is currently uncertain whether international institutions can be as *input-legitimate* as national institutions. The argument is not that international institutions are inherently incapable of being more input-legitimate than national institutions. The point is merely that current political practices at the international level can be vulnerable to strong transnational actors that represent corporate economic interests rather than public and environmental interests (Erman 2010; Uhlin 2010; Lahson 2005).

On the other hand, we have seen that national democracies are also increasingly vulnerable to elitism and centralised decision-making processes. In this regard, international institutions may balance some of the malfunctions at the national level by, for example, thwarting or preventing powerful unjust actors from abusing their power to disadvantage the weak and poor and empowering vulnerable actors to protect their fundamental interests (Caney 2006, 742).

The third criterion for determining the level where democracies might be most effective is equal distribution of benefits. *The tragedy of the few* story reveals that there is currently a strong corporate centralisation of the access to natural resources. This centralisation is allowed in large part by national and international political agreements in which private corporations are licensed to subtract and sell the resources. As argued in Chapter 5, these private or state-based corporations do not have full property rights to the earth's natural resources. Thus, a fair question is who benefits from the subtraction of natural resources and moreover, who suffers from the negative side effects of, for example, subtracting oil that pollutes the air with carbon. Is there a fair distribution of benefits and burdens (Caney 2011, 526)?

Indeed, the fact is that corporations benefit financially disproportionately more than ordinary citizens suffer and will suffer from carbon pollution. Whereas taxation might remedy some of the unfair distribution of benefits, the tax remedy is not sufficient because it is necessary to invoke the policy options outlined in Chapter 5, which suggested *banning* subtraction of certain types of resources and *distributing* the access to subtraction more equally as possible strategies for realising a fair distribution of benefits and burdens.

National and global democracy

This section is concerned with the relationship between the instrumental and teleological account of collective ends that the democratic institutions are obliged to fulfil and what level of democratic institutionalisation should be demanded. The problem stems from the fact that the quality and strength of many national democracies has deteriorated. In addition to the democratic problems at the national level, there are several democratic problems attached to the global level of governance as well. In particular, global politics lacks a demos, accountability, democratic values, and formal access to decision making (Erman 2010, 173–180). This has happened whilst the economy has been globalised through the liberalisation of trade, the deregulation of financial markets, and the privatisation of state assets (Higgott and Erman 2010, 449). The current result is what Richard Higgott and Eva Erman have called an 'over-developed global economy' and an 'underdeveloped global polity' (Higgott and Erman 2010, 452).

The normative argument for at what level democratic climate governance should be implemented is situated in the intersection of two profoundly destabilising changes in the way we view the world: one is economic; the other is political. On the political front, the shift is from the authority of the nation-states towards a fracturing of state sovereignty with transnational financial and labour mobility that puts great pressure on the current forms of democratic governance. The 'old' ways of understanding politics centred around democratically legitimised decision making is now confronting a 'new' politics of pluralism, multi-level stakeholders, the power of multinational corporations, and fragmentation of political authority (Jasanoff 2005, 14). In particular, the transnationalism of contemporary politics has given rise to a more elitist approach to democratic politics that is dominated by national and international networks of political and economic elites (Scharpf 2009, 14; Buch-Hansen and Wigger 2011, 139).

On the economic front, the current shift is towards a greater influx of economic thinking and calculation into political and ethical reasoning. Whereas a Marxist, of course, would argue that politics and economics always have been closely linked, it has been quintessential for political philosophy and democratic theory to believe that politics is not dominated by economic calculation but by democratic values, people's legitimate preferences, and public reason. However, it has become harder to maintain this approach in recent years. Modern climate politics is highly complex and unsuitable in many ways for democracy decision making and public reasoning. Take the example of the global economic instruments that have been developed to mitigate CO_2-emissions: because they are very complex, few people know or even understand the details, let alone all of their implications (ICLEI 2010, 20).

Several suggestions have been made to remedy the democratic deficit. Most prominent are the normative arguments for strengthening global civil societal deliberation, for example, by including NGOs in international UN decision processes (Dryzek and Stevenson 2011) and by institutionalising and constitutionalising formal access and equal participation in decision-making procedures (Erman 2010, 190). A good example of the latter is the strengthening of the European Parliament's influence, access, and participation in European legislation in the 2009 Treaty of Lisbon.

Polycentric democracy

Until now an underlying assumption of this discussion has been the political relevance of distinguishing between national and global policy institutions. As elaborated on in Chapter 2, one important debate in the global justice literature concerns national obligations versus global obligations. I opted for a split-level approach in which moral obligations are universal whilst the level of institutionalisation of moral principles is contingent on what political contexts are perceived to be significant. The notion of political contexts lies at the core of the fact-sensitive account of normativity. As such, the comparable moral significance of political contexts depends on factual circumstances. When we engage in normative inquiries of the institutionalisation of access policies, we need to address whether the conclusions we make concern both national or global political contexts or only one of them.

Some theorists have sought to identity politically relevant contexts with a particular culture, for example, by assuming that cultures are bound to national identities (D. Miller 2007), whereas others assume that cultures have a global reach (Benhabib 2004, 45). Indeed, in the cosmopolitan literature, distinguishing between *thick* and *thin* cultural communities is widely accepted. A thick cultural community may be the Amish people, for instance, whilst an example of thin cultural communities could be an international group of football fans or Facebook friends. Another way of suggesting that we are living in a global community is to draw attention to 'the dense thicket of rules that sustain our life together [which] includes some of the most mundane things imaginable: postal and telephone conventions, airline safety and navigation standards, the law of international trade [. . .] and so on' (Waldron 2006, 83).

However, the mere existence of cultural affiliations and international rules and norms does constitute an argument for what political contexts and institutions should be weighed. Consider the Amish group, which has established a strong internal cultural affiliation that we can assume is stronger than most national identities. Nonetheless, the Amish group is not considered a more important political level than the national level of the country where it resides. Thus, I conclude that cultural affiliations cannot tell us anything about what political context and institutions are significant. Likewise, the cases of international standards, norms, and conventions on the one hand and international groups of sports fans on the other say little about the relevance of political institutions with regard to climate governance.

In contrast, recent developments on the political and economic front make it less relevant to establish whether the national or global level of politics is of particular importance. As we have seen, the supra-national parliament of Europe and the subnational management of big cities are important democratic venues of climate governance that are not captured by the dichotomy between national and global contexts. Another layer of policy governance that trespasses the national and global context is international organisations such as the OECD, WTO, IMF, and the World Bank. These institutions have a great impact on lower levels of politics and their capacity to do climate politics, but they are incomparable with, for example, European and national polities because of their issue-specific portfolios.

One way to comprehend the international landscape of institutions has been provided by James Bohman, who argued that democracy across traditional national borders should be considered as *polities of demoi*. Standard democratic theory organises parliamentary institutions around one demo. In contrast, the notion of *the polity of demoi* reflects the fact that democratic decisions across borders are not merely an aggregate of peoples' interests or visions but a new multifaceted structure in which distinct institutions have overlapping competencies distributed at different policy levels (Bohman 2010, 10).

Thus, given the complexity of modern political governance, I take it to be impossible to choose one institution that should be prioritised over others. The fact-sensitive account of normative principles is not a detailed recipe for specific climate action and institutions: I take it to be a reflection of concern that democratic decision processes prescribe specific ends, means, and levels of institutionalisation. However, the fact-sensitive account can clarify normative principles and problem-specific criteria for governance, including, for example, the three access criteria of climate governance. Whether the access criteria should be implemented at the local, national, or global level is a contingent matter that requires thorough knowledge about the factual circumstances of politics. We should also keep in mind that factual circumstances are unstable conditions that could potentially make things possible tomorrow that were impossible yesterday.

Notwithstanding the inevitable caveats that accompany such details, one general point can be stressed. Given the complexity and multi-level stakeholder policy challenges the world is currently facing, a reasonable suggestion is to embrace a polycentric combination of local, national, and international policy levels, which provides a mixed system in which both procedural and instrumental concerns are satisfied (Caney 2006; Ostrom 2012). Such a system has the potential to construct what we could call a *vertical check-and-balance* system of the policy levels inbetween. More generally, instead of focusing on the conflicts between specific political contexts, a fruitful approach would be to emphasise what Christopher K. Ansell has called 'evolutionary learning' between local, national, and global levels. Evolutionary learning is produced by shifting back and forth between local and cosmopolitan standpoints, which generates an exchange of local and cosmopolitan perspectives (Ansell 2012, 13).

Here, the civic, political, and institutional exchange between the various political contexts is stressed, and thus the ability which individuals, group agents, and communities have to improve their knowledge and problem-solving capacity over time through continuous inquiry, reflection, deliberation, and experimentation

(Ansell 2012, 5). In other words, when we focus on 'the impact of knowledge-ability', the capacity of citizens to govern themselves is stressed, as is their ability to enhance the conditions for democracy at local, national, and international levels (Stehr 2008, 5–6). The normative arguments for improvement of national democracies are compatible with the argument for democratisation of international organisations. In fact, I take democratisation at the lower and higher levels of politics to be mutually reinforcing. So conceived, one of the main challenges is not to create one global demos but to establish an institutional structure which, as Bohman has formulated it, allows for the reflexive and democratic capacity of its citizens 'to initiate legitimate democratic reforms' (Bohman 2010, 10).

In order to improve the exchange, reflection, and deliberation between the various political contexts, several approaches are worth considering. One includes bottom-up, disaggregated, and polycentric solutions to climate action (Prins and Rayner 2007; Stehr and Storch 2009; Ostrom 2012; Stehr 2013). Here, the main goal of global climate governance is, for example, not one single global and legally binding agreement or a new Kyoto Protocol established by elitist political bodies with low epistemic and deliberative value. Instead, the solutions to the failures of climate governance should be found in locally fostered innovative initiatives that result in overlapping policies at city, subnational, national, and international levels (Prins and Rayner 2007; Ostrom 2012; Bulkeley 2013).

Another suggestion focuses on the improvement of democratic deliberation in policy institutions (Dryzek and Stevenson 2011; Baber and Bartlett 2005) and the direct influence on the decision processes in existing local, national, and international parliaments or new institutions, such as what some have called 'directly deliberative polyarchy' (Cohen 1997). For example, a democratisation of the WTO would increase the capacity of lower levels of climate governance since the WTO constrains some environmental regulations (Singer 2002). The IMF also recently called for 'improvements of transparency' and 'dissemination of information' on, for example, 'the magnitude of oil and coal subsidies' (IMF 2013, 1).

The point of the first two arguments for democratic reforms is the idea that effective democratic climate politics enables a better utilisation of the potential benefits of democratic institutional design by allowing citizens, stakeholders, and politicians to contribute collectively to innovative solutions. As we saw earlier, a premise for the epistemic gains of democracy is that the citizens are *independent* and *competent*. The third argument for transparency touches upon this challenge. Remember, competence does not refer to the level of education but to the 'truth bias' of people's judgements. Here, an important premise is that people have access to the necessary information about climate and energy policies. Hence, in order to provide democratic reforms, one condition is to empower citizens' independence and competence. Only in this way is it possible to enhance democracies' control over policy options and thereby consider democracies as institutionalised collective agents that have collective responsibility for climate change politics.

On the teleological account of collective ends and climate change, it is quintessential that these political institutions are democratic ones – independent of the level of climate governance. Thus, the occupants of institutionalised positions in democracies are not merely responsible for fulfilling the collective ends related to climate governance but also to maintain and, if necessary, reform the democratic quality at all political levels. By so arguing, the occupants of institutionalised positions are responsible for upholding an institutional design in which the capacity of the local and national democracies is not undermined.

Conclusion

In this chapter, I have argued that occupants of institutional positions should be held collectively responsible for choosing a too narrow policy mix that neglects the problems related to *the tragedy of the few* and thus the negative side effects on reversing climate politics at a later point. They should not primarily be held collectively responsible for not doing enough now but rather for not providing the basis for doing the right thing at a later point. Moreover, they should be held collectively responsible for not promoting democratic reforms at all politically relevant levels.

Note

1 The multi-interdisciplinary research approach which allows for a stronger representation of the social sciences in climate science has also been requested by IPCC panels (IPCC 2013). Moreover, the multi-interdisciplinary research approach fits well with the objective of many current research policies in which interdisciplinary collaboration is advanced (Holm et al. 2012; *Visions for Horizon 2020* 2012).

References

Ansell, C. K., 2012. *Pragmatist democracy: evolutionary learning as public philosophy.* Oxford: Oxford University Press.

Baber, W. F. and Bartlett, R. V., 2005. *Deliberative environmental politics: democracy and ecological rationality.* Cambridge, MA: MIT Press.

Benhabib, S., 2004. *The rights of others: aliens, residents and citizens.* Cambridge: Cambridge University Press.

Bjurström, A. and Polk, M., 2011. Physical and economic bias in climate change research: a scientometric study of the IPCC Third Assessment Report. *Climatic Change*, 108 (1–2), 1–22.

Blyth, M., 2013. Paradigms and paradox: the politics of economic ideas in two moments of crisis. *Governance*, 26 (2), 197–215.

Bohman, J., 2010. Introducing democracy across borders: from dêmos to dêmoi. *Ethics and Global Politics*, 12 (1), 1–11.

Buch-Hansen, H. and Wigger, A., 2011. *The politics of European competition regulation: a critical political economy perspective.* New York, NY: Routledge.

Bulkeley, H., 2013. *Cities and climate change.* New York, NY: Routledge.

Caney, S., 2006. Cosmopolitan justice and institutional design: an egalitarian liberal conception of global governance. *Social Theory and Practice*, 32 (4), 725–756.

Caney, S., 2011. Humanity, associations, and global justice: in defence of humanity-centred cosmopolitan egalitarianism. *The Monist*, 94 (4), 506–534.

Cohen, G. A., 1997. Where the action is: on the site of distributive justice. *Philosophy and Public Affairs*, 26 (1) (Winter), 3–20.

Dryzek, J. S. and Stevenson, H., 2011. Global democracy and earth system governance. *Ecological Economics*, 70 (11), 1865–1874.

Erman, E., 2010. Why adding democratic values is not enough for global democracy. *In:* E. Erman and A. Uhlin, eds. *Legitimacy beyond the state? Re-examining the democratic credentials of transnational actors*. New York, NY: Palgrave Macmillan, 173–192.

Fukuyama, F., 2004. *State-building: governance and world order in the 21st century*. Ithaca, NY: Cornell University Press.

Higgott, R. and Erman, E., 2010. Deliberative global governance and the question of legitimacy: what can we learn from the WTO? *Review of International Studies*, 36 (2), 449–470.

Holm, P., Goodsite, M., Cloetingh, S., Agnoletti, M., Moldan, B., Lang, D. J., Leemans, R., Oerstroem, J., Moeller, Pardo Buendía, M., Pohl, W., Scholz, R. W., Sors, A., Vanheusde, B., Yusoff, K., and Zondervan, R., 2012. Collaboration between the natural, social and human sciences in Global Change Research. *Environmental Science & Policy*, 28, 25–35.

Howard-Snyder, F., 2006. 'Cannot' implies 'not ought'. *Philosophical Studies*, 130 (2), 233–246.

ICLEI Global Reports, 2010. *Cities in a post-2012 climate policy framework. Climate financing for city development? Views from local governments, experts and businesses*. Available from: www.iclei.org/fileadmin/PUBLICATIONS/Papers/Cities_in_a_Post-2012_Policy_FrameworkClimate_Financing_for_City_Development_ICLEI_2010.pdf [Accessed 15 January 2017].

IMF (International Monetary Fund), 2013. *Energy subsidy reform: lessons and implications*. Available from: www.imf.org/external/np/pp/eng/2013/012813.pdf [Accessed 10 January 2017]. IPCC (Intergovernmental Panel on Climate Change), 2013. *Summary for policymakers*. Twelfth Session of Working Group I Approved Summary for Policymakers. Available from: www.climatechange2013.org/images/uploads/WGIAR5-SPM_Approved27Sep2013.pdf.

Jagers, S. C. and Duus-Otterström, G., 2008. Dual climate change responsibility: on moral divergences between mitigation and adaptation. *Environmental Politics*, 17 (4), 576–591.

Jasanoff, S., 2005. *Designs on nature: science and democracy in Europe and the United States*. Princeton, NJ: Princeton University Press.

Kaul, I. and Mendoza, R. U., 2002. Advancing the concept of public goods. *In:* R. A. Musgrave and P. B. Musgrave, eds. *Providing global public goods*. Oxford: Oxford University Press, 125–143.

Lahson, M., 2005. Technocracy, democracy, and U.S. climate politics: the need for demarcations. *Science, Technology & Human Values*, 30 (1), 137–169.

List, C. and Pettit, P., 2011. *Group agency: the possibility, design, and status of corporate agents*. Oxford: Oxford University Press.

Miller, D., 2007. *National responsibility and global justice*. Oxford: Oxford University Press.

Miller, S., 2010. *The moral foundations of social institutions: a philosophical study*. Cambridge: Cambridge University Press.

OECD (Organization for Economic Co-operation and Development), 1992. *General Distribution OCDE/GD (92) 81. The polluter-pays principle: OECD analyses and recommendations*. Paris: OECD.

Ostrom, E., 2012. Green from the grassroots. *Project Syndicate*, 12, 1–3.

Ostrom, E., Gardner, R., and Walker, J., 1994. *Rules, games, and common-pool resources*. Ann Arbor, MI: University of Michigan Press.

Peck, J., 2010. *Constructions of neoliberal reason*. New York, NY: Oxford University Press.

Plant, R., 2009. *The neoliberal state*. Oxford: Oxford University Press.

Prins, G. and Caine, M. E., 2013. *The vital spark. Innovating clean and affordable energy for all*. The Third Hartwell Paper, July. London: LSE Academic Publishing.

Prins, G. and Rayner, S., 2007. *The wrong trousers: radically rethinking climate policy*. Available from: eureka.sbs.ox.ac.uk/66/ [Accessed 26 August 2017].

Rostbøll, C., 2009. *Deliberative freedom: Deliberative democracy as critical theory*. Albany, NY: State University of New York Press.

Sagoff, M., 2008. *Philosophy, law and the environment*. New York, NY: Cambridge University Press.

Scharpf, F., 2006. The joint-decision trap revisited. *Journal of Common Market Studies*, 44 (4), 845–864.

Scharpf, F., 2009. Legitimacy in the multilevel European polity. *European Political Science Review*, 1 (2), 173–204.

Schlager, E., Blomquist, W., and Yan Tang, S., 1994. Mobile flows, storage, and self-organized institutions for governing common-pool resources. *Land Economics*, 70 (3), 294–317.

Singer, P., 2002. *One world*. New Haven, CT: Yale University Press.

Stehr, N., 2008. Introduction: is freedom a daughter of knowledge? *In:* N. Stehr, ed. *Knowledge and democracy: a 21st century perspective*. New Brunswick, NJ: Transaction Publishers, 1–8.

Stehr, N., 2013. An inconvenient democracy: knowledge and climate change. *Society*, 50, 55–60.

Stehr, N. and Von Storch, H., 2009. Climate protection. *Journal of Consumer Protection and Food Safety*, 4, 56–60.

Uhlin, A., 2010. Democratic legitimacy of transnational actors: mapping out the conceptual terrain. *In:* E. Erman and A. Uhlin, eds. *Legitimacy beyond the state? Re-examining the democratic credentials of transnational actors*. New York, NY: Palgrave Macmillan, 16–37.

Visions for Horizon 2020, 2012. *Copenhagen research forum*. Available from: www.crf2012.org/upload/dtu%20kommunikation/crf_rapport_rgb_singlepage.pdf [Accessed 16 February 2017].

Waldron, J., 2006. Cosmopolitan norms. *In:* S. Benhabib, ed. *Another cosmopolitanism*. Oxford: Oxford University Press, 13–44.

Conclusion

The aim of this book was to find a way of thinking about moral responsibility for climate change that is not too demanding but still ambitious. I began this discussion by outlining why climate change is morally wrong. I suggested that in our response to climate change, we must not only consider moral wrongdoing: we should also ask questions about how and why climate change is happening. This differs significantly from the economic approach, which draws attention to the costs of mitigating climate change; by contrast, the moral philosophical account of climate ethics is concerned with who should bear the costs and, moreover, the natural scientific focus on the end results of climate change. Together these three disciplines converge in neglecting the social and political theoretical questions of who are the relevant agents, to whom it is fair to assign responsibility, and what explanatory models are available to us in order to comprehend why contemporary democratic climate change fails.

In order to circumvent this focus on costs and emissions, I have suggested a fact-sensitive approach to normative political theory. My main argument in invoking this approach has been that concrete ought-assignments, such as responsibility for climate change, require a fact-sensitive theorisation that includes a consistent and normatively robust combination of normative thinking and methodologically embedded notions on agency, rationality, and explanation.

I argued in Chapter 1 for distinguishing between abstract and concrete normative principles, with the latter being neither realist nor non-ideal. Instead, they are fact-sensitive principles that are warranted in abstract principles but concern rules of regulation, institutionalisation of principles, political contexts, and concrete ought-assignments. In Chapter 2, I elaborated on this account by arguing that abstract principles fail to determine these concrete matters whilst the concrete principles in the split-level model are normatively robust and factually relevant.

In this account, the goal of Chapter 3 was to remedy the under-theorisation of what facts are significant in the determination of concrete ought-assignments. I drew attention to a number of relevant constraints but focused on the so-called soft *internal* and *external* constraints, arguing that these can- and fitness-conditions for agency feed into the normative discussion of who ought to do what. In invoking these meta-theoretical questions in Part 1, the way was paved for engaging in two of the key ideas in the book: to reconceptualise what the political problems of climate change are and to reconceptualise who are the relevant moral agents in this issue on the one hand, as well as to reframe who it is fair to assign moral responsibility to on the other.

In Part 2, I was primarily occupied with the investigation of whether the methodologies of rational-individualism and non-reductionist institutionalism provide a framework within which agents are fit to be held responsible according to the three soft external criteria of fitness-conditions: normatively significant choice, sufficient knowledge, and control over the choice. I concluded that the rational-individualistic framework, which dominates many climate mitigation policies, contains no moral agent to solve the economic incentive problems politically by virtue of the ways that agency is comprehended in *the tragedy of the commons* story. By so arguing, I emphasised in Chapter 6 that it is apprehensible that climate scholars and activists argue in favour of radical policy fixes of the incentive problems, by, for example, embracing bioengineering and geo-engineering on the one hand and market-based or authoritarian policies on the other. These policy suggestions, albeit very different in character, converge in the critique of political and democratic climate governance.

As a way of re-joining the critique of democratic climate governance, I argued that *the tragedy of the commons* is an inadequate story that fails to help us understand the full complexity of the political problems related to climate change, which cannot be reduced to a question of lack of economic incentive. In contrast, I suggested *the tragedy of the few* as a way to address the unequal access to resources, benefits, and decision processes from which climate governance deficiencies suffer. In this account, the main problem of climate change is not too much democratic governance but that the existent quality and capacity of national and global democratic politics are deteriorating whilst being increasingly dominated by elitism, technocracy, expert groups, and corporate interests. In this environment, the cause of environmental harm and climate change to which *the tragedy of the few* story refers, remains unaddressed: that owners of land and technological equipment have a privileged right to subtract and use resources to which the agent has no full ownership.

Normally, we tend to avoid a political theory that comments on whether one type of politics or another ought to be addressed: we accept that the exercise of politics is the job of democratic policymakers. Nonetheless, by applying the fact-sensitive account of normativity, I allowed for normative investigations of the kind: *if Q, it is good that X*, in which a normative valuation of specific policies can be warranted in fundamental principles. The trick is that it is not the fundamental principles that evaluate the specific policies that would make most policies unwarranted. The idea instead is to construct fact-sensitive normative inquiries in their own right. Consider the examples of access that I have discussed. The normative challenge of access to natural resources, benefits, and decision processes occurs because of the conflicts between territorial rights and basic rights to life, health, and subsistence. I have argued that given that we assume a set of basic rights, we can say that the politics that address the access problems are better than the politics that disregard these problems. And, arguably, we might suggest that the relevant moral agents *should* choose the former politics rather than the latter.

The idea is not that a political theoretical account of climate change should replace the democratic decision-making processes. On the contrary, the key idea is that certain institutionalised group agents are obliged to fulfil their institutional duties – that is, to provide the collective ends and purposes that define the institution. Addressing the access problems that are related to *the tragedy of the few* are part and parcel of the *raison d'être* of most contemporary political institutions at local, national, and international levels. One important method to improve the likelihood that occupants of institutionalised positions will satisfy the collective ends they are obliged to fulfil is to democratise all levels of political institutions. Well-functioning democratic institutions in which participants are independent and competent interlocutors increase the epistemic quality of the political process and thereby the awareness and knowledge about climate change and related access problems. Moreover, I contend that the fitness of democratic institutions to be held morally responsible for the governance deficiencies is likewise increased.

Because one of the goals of Part 3 was to investigate whether a group of agents in democratic institutions should be held morally responsible for climate change, I examined the group of democratic citizens and occupants of institutionalised positions. Whilst many scholars have focused on the individual responsibility of rational agents, I argue that rational agents are not moral agents with regard to climate change but are moral agents with regard to environmental harm. Moreover, the shift of methodological framework to the non-reductionist institutionalism model allows us to conclude that democratic citizens are reasonable agents who should be held morally responsible for not doing more jointly to promote green government and the democratisation of contemporary elitist policymaking processes.

However, these individual agents should not be morally responsible *as individuals* because it is unfair to assign moral responsibility to citizens who do not intend the bad outcome to happen. But these citizens nonetheless contribute to the bad outcome because of the surrounding social and political structures that make it impossible or at least very difficult to avoid the bad outcome.

Thus, instead of focusing only on the level of individual and joint responsibility, more attention should be paid to the collective responsibility of occupants of institutionalised positions in democratic societies. It is fair to assign them responsibility for two reasons: first, their moral deficiencies cannot be morally excused since collective responsibility is defined by consequentialist remedial responsibility. Second, it is fair to assign collective responsibility to occupants of institutionalised positions because their roles are defined by remedying unintended and negative effects that are caused jointly by all participants. Furthermore, they are not only responsible for not choosing the best climate politics now but also for not providing and sustaining the democratic quality and capacity of the political institutions that would have made it easier to shift policy framework on climate policy.

I conclude on a note of democratic governance. If we believe that climate change is morally wrong and that something ought to be done about it, the best option is *not* to push for ambitious mitigation policies that do not fit the current institutional capacity of our current systems of government, as was the case with the Kyoto Protocol. Instead, we need to take a step back and reconstruct the institutional capacity of all relevant political levels. We need to increase the opportunities of direct democratic influence on policies conducted by international organisations, such as the WTO, and national and regional governmental bodies, such as national parliaments and the European Union. Our goal should be to enhance the transparency of climate change decision making and improve free access for all to relevant knowledge about the issue – that is, free access to education, science, and public media.

Index

Note: Page numbers in italics indicate figure and in bold indicate tables on the corresponding pages.

2013 Climate Action Plan 97

abstraction 21–22
access: to knowledge 103–104; policy of 101–103
active rights holders 148
acts and omissions, asymmetry between 149–151
agency, group 60–62
Ansell, C. K. 171
applied ethics 9
approximate moral cosmopolitanism 33
authoritarian solution 121–122

backward-looking responsibility 135
Bader, V. 43
Barry, B. 85
behavioural space 58, 74
benevolent dictator 122
bioengineering 119–120
blame criteria 136–138
Bohman, J. 171, 172
bounded rationality 77
Brown, G. W. 34

can- and *fitness*-conditions 10, 48–49, 53–56, **54**; *see also* fact-sensitive ought-assignments
Caney, S. 98
chlorofluorocarbons (CFCs) 2
climate change: environmental harm and 88–89; fact-sensitivity and normativity regarding 8–10; idealism and realism regarding 7–8; moral responsibility for 4–7, 176–179; political failures of actions on 1–2; reframing of political

challenge of 97–98; responsibility to address 2–4; sceptics on 13n5
climate science responsibility 164–166
cognitive capacities 117–118
Cohen, G. A. 25–26, 28–29
coherentism 24–25
collective benefits ignorance and the market 106–109
collective ends 158–160; democracy and 167–169
collective institutional duties 157–158
collective responsibility 133–134; blame criteria and 136–138; climate science responsibility and 164–166; collective ends and 158–160, 167–169; collective institutional duties and 157–158; democratic climate governance and 138–139; dynamic duties and 160–164; joint and 134–136; national and global democracy and 169–170; polycentric democracy and 170–173; teleological account of democracy and 166–167
common, but differentiated responsibility 3
common-pool resources 99
concrete and abstract principles 21–23; fact-sensitive ought-assignments and 48–49; in moral cosmopolitanism 33; non-ideal account and 37–38
Conference of the Parties (COP) 15 2, 13n3
constraints, soft 56–58
contextualist model 38–40
control 115–116; *see also* democratic climate governance
Copenhagen Accord 117, 127–128n6
cosmopolitanism: contextualist model and 38–40; indeterminacy challenge and

35–37; institutional 45n4; interactional 45n4; levels of institutionalisation and 43–45; liberal nationalism and 34–35; moral 32–34; non-ideal account and 37–38; split-level model and 40–43
Council Recommendation on Guiding Principles concerning International Economic Aspects of Environmental Policies 87
cultural affiliations 170

democracy: collective ends and 167–169; national and global 169–170; polycentric 170–173
democratic citizens 73, 141–142; asymmetry between acts and omissions and 149–151; joint responsibility and 152–153; moral deficit and 142–144; negative and positive duties and 148–149; perfect and imperfect duties and 151–152; right-libertarian moral deficit argument and 144–148, 154n5
democratic climate governance 109–110, 115–116, 126–127; authoritarian solution in 121–122; cognitive and moral capacities and 117; collective responsibility and 138–139; democratic reforms and 124–126; factual aspects of critique of 122–124; neoliberal policy proposals in 116–117, 161–163; proactionary policies in 116, 118–121, 128n10, 161–163
democratic reforms 124–126
descriptive component of moral responsibility 6
descriptive judgements 23
diachronic aspect of responsibility 153
Dietz, T. 83
directly deliberative polyarchy 172
disagreements about facts 28–30
discount rate 74–75
dynamic duties 160–164

ecofeminism 128n7
emergentism 61
environmental harm and climate change 88–89
equal right to access 102
Erman, E. 169
European Commission 87, 136
European Union 40, 116–117, 179
extended producer responsibility (EPR) 92n26

fact-sensitive ought-assignments 48–49; group agency and 60–62; methodological assumptions 58–60; moral premises 62–64; ought-entailments 52–53; *ought implies can* (OIC) 50–52; ought-judgements and 49–50; soft constraints 56–58; what can 'A' do? 53–56, **54**
fact-sensitive political theory 26–27
fact-sensitivity 8–10; concrete and abstract principles 21–23, **22**; disagreements about facts and 28–30; facts and principles in 23–24; in political theory 26–27; relational value judgements and 24–26
Fahlquist, J. 145
Faustian bargain 164–165
feasibility constraints 9
first-order responsibility 143
fitness-conditions: group agents 95–112; rational agency 73–92
forward-looking responsibility 135
France as flawed democracy 128n12
Fukuyama, F. 162
full ownership rights 101
fundamental value judgements 24–25

Galton, F. 109
Geuss, R. 8
Gilabert, P. 55, 57, 146–147
'good authoritarianism' 121–122
Green Climate Fund 117
greenhouse gases (GHG) 2, 12n2, 87
green virtues 80, 111–112n10
group agency 60–62; access to knowledge and 103–104; democratic climate governance and 109–110; market and collective benefits ignorance and 106–109; market and values and 104–106; natural resources and human rights and 98–100; policy of access and 101–103; tragedy of the few and 95–98; two sets of rights and 100–101

Hardin, G. 76–77
Hayek, F. 104–109, 117
Heede, R. 96
Held, D. 33, 34, 35, 36–37
Higgott, R. 169
Holtug, N. 35
Howard-Snyder, F. 56, 57
human rights and natural resources 98–100
Hurley, S. 75

idealism 7–8
incentive critique 115, 117, 118, 121–122
indeterminacy challenge 35–37;
    contextualist model and 38–40; levels of
    institutionalisation and 43–45; non-ideal
    account and 37–38; split-level model
    and 40–43
individualism 32–33, 59, 90–91n7
individual-rationalism theories 59
information pooling 110
input-legitimate institutions 168
institutional cosmopolitanism 45n4
institutionalisation, levels of 43–45
institutionalism, non-reductionist
    methodology of 59–60, 103
interactional cosmopolitanism 45n4
International Monetary Fund (IMF) 97,
    116, 171, 172
International Panel for Climate Change
    (IPCC) 2
irrational rational behaviour 76

Jamieson, D. 80
Johnston, D. 116
joint ownership 111n6
joint responsibility 134–136, 152–153
judgemental capacity 58
justice, conceptualisation of 41
justification and formulation of moral
    principles 39–40

knowledge deficit 142
Kyoto Protocol 2, 3, 117, 172, 179

Lawford-Smith, H. 55, 57
layered cosmopolitanism 33
Le Grand, J. 125
Liao, S. M. 120, 124, 128n8
liberalism 29, 123, 127n2
liberal nationalism 44, 45;
    cosmopolitanism and 34–35; split-level
    model and 41–43
Lichtenberg, J. 147, 148
List, C. 58, 62
locked-in syndrome 143
Logic of Scientific Discovery 112n11
Lovelock, J. 122
low-capacity challenge 143

market, the: collective benefits ignorance
    and 106–109; values and 104–106
Mason, A. 27
medical moral enhancement 119
meritocratic elite 122

Midlarsky, M. 123
Miller, D.: on cosmopolitanism 34–35,
    38–44; on *ought implies can* (OIC) 51
Miller, S. 62, 158–159
miscalculation of social benefits 81–84
moderate cosmopolitanism 33; non-ideal
    account and 37–38
Montreal Protocol 2, 12n1
moral capacities 117–118
moral cosmopolitanism 32–34;
    indeterminacy challenge and 35–37;
    liberal nationalism and 34–35; non-ideal
    account and 37–38
moral deficit and democratic citizens
    142–144; right-libertarian moral deficit
    argument 144–148
moral imperative 151–152
moral motivation 118
moral premises 62–64
moral responsibility 4–7, 176–179;
    collective (*see* collective responsibility);
    idealism and realism and 7–8; moral
    cosmopolitanism and 32–34
moral threshold 98
moral universalism 32–33

national and global democracy 169–170
nationalism, liberal 44, 45;
    cosmopolitanism and 34–35; split-level
    model and 41–43
natural resources and human rights 98–100
negative duties 148–149; are not enough
    argument 146–148
neoliberalism 116
neoliberal policy proposals 116–117,
    161–163
neurotransmitters 119
no-impact challenge 143
nomocratic approach 160, 166–167
'non-culpable ignorance' 142–143
non-divided resources 100–101
non-fundamental value judgements 24–25
non-ideal account 37–38
non-idealism 9
non-reductionist methodology of
    institutionalism 59–60, 103
normative component of moral
    responsibility 6
normative judgements 23
normative significance 58
normativity 8–10; four types of 22, **22**
norms, social 74, 77, 79–81
Nozick, R. 117
Nussbaum, M. 33, 35, 36–37

Obama, B. 97
omissions 149–151
O'Neill, O. 9, 21–22, 36–37
Ophuls, W. 122
Organisation for Economic Co-operation
    and Development (OECD) 87–88, 116,
    171
Ostrom, E. 77, 82, 83, 84
ought-entailments 52–53
ought implies can (OIC) 50–52; collective
    responsibility and 133–134, 158; moral
    premises 62–64; ought-entailments and
    52–53
ought-judgements 49–50
output legitimacy 168

Parfit, D. 75–76
Paris Agreement 2
passive rights holders 148
perfect and imperfect duties 151–152
Persson, I. 120, 128n9
Pettit, P. 58, 62
Plant, R. 162
Pogge, T. 36, 41
political challenge of climate change,
    reframing of 97–98
political failures of climate actions 1–2
'political philosophy for Earthlings' 34–35
political theory, fact-sensitive 26–27
polities of demoi 171
polluter pays principles (PPP) 58, 86–88;
    environmental harm and climate change
    and 88–89
polycentric democracy 170–173
Popper, K. 112n11
positive duties 148–149
practice-dependent thesis 39
prisoner's dilemma 59, 90n5
private goods 99
proactionary policy proposals 116, 118–121,
    128n10, 161–163
public goods 99
Putnam, H. 23

Rachels, J. 150
rational agency, fitness-conditions of 73, 89;
    environmental harm and climate change
    and 88–89; miscalculation of social
    benefits 81–84; self-interest theory 74–76;
    snatch dilemmas 77–79, **78**, **79**; social
    behavioural spaces 84–86; social context
    and norms 79–81; social dilemma theory
    76–77; and who are the polluters? 86–88
rational agents 73

rational choice institutionalism (RCI) 90n6
rational choice theory 59, 74, 81, 84–85
rationality, bounded 77
Rawls, J. 28–29
realism 7–8
REDD+ programme 117
reforms, democratic 124–126
relational value judgements 24–26
relevant control 58
repeated many-person dilemmas 77
right-libertarian moral deficit argument
    144–148, 154n5
right to deplete, exploit, or pollute 103
right to use 101
Road to Serfdom, The 108
Robock, A. 120–121
Rose, G. L. 1

Saad-Filho, A. 116
Sangiovanni, A. 38, 39, 40
Savulescu, J. 120, 124, 128n9
Schomberg, R. 102
Science 76
Searle, J. 106
self-imposed inability 56
self-interest theory 74–76, 90n2
Shearman, D. 123–124
Shue, H. 86, 142–143
Simon, H. A. 77
Singer, P. 149
single-level methodologies of
    individualism 59
Sinnott-Armstrong, W. 57, 143
Smith, J. W. 123–124
snatch dilemmas 77–79, **78**, **79**
social behavioural spaces 84–86
social benefits, miscalculation of 81–84
social context and norms 79–81
social dilemma theory 76–77
social norms 74, 77; social context and
    79–81
soft constraints 56–58, 141
split-level model 40–41; liberal
    nationalism and 41–43
stance-theory 165–166
Starship Enterprise view 38–39
Stehr, N. 164–165
Stern, N. 59
Stern, P. C. 83
Stern, R. 52–53
Stern Review, The 59
supervenience 61–62
systematic miscalculation of collective
    benefits 82–83

Tan, K. 36
teleological and maximal theory of
   democracy 139, 166–167
teleological property 160
teritorially divided resources 100–101
Theory of Justice, A 28
thick cultural communities 170
thin cultural communities 170
tragedy of the commons 59, 76, 78, 81,
   104; climate science responsibility
   and 165; collective responsibility and
   159–160; democratic climate governance
   and 115, 117, 123; versus tragedy of the
   few 95–96, 101
tragedy of the few 95–98, 115, 117, 168;
   climate science responsibility and 165;
   collective responsibility and 159–160;
   democratic climate governance and 123,
   124, 125, 126, 139; equal right to access

and 102; human rights and 98–99; two
   sets of rights and 100–101
transnationalism with cosmopolitan
   inflection 33

United Nations 2, 87
universalism, moral 32–33
utilitarianism 137

values and the market 104–106; objective 107
vertical check-and-balance system 171
voluntarism 121, 122, 144–145
Von Storch, H. 165

Williams, B. 7–8
World Bank 87, 116, 171
world ownership 111n6
World Trade Organisation (WTO) 33–34,
   40, 116, 171, 172, 179